Semisolvability of Semisimple
Hopf Algebras of Low Dimension

of the
American Mathematical Society

Number 874

Semisolvability of Semisimple Hopf Algebras of Low Dimension

Sonia Natale

March 2007 • Volume 186 • Number 874 (fourth of five numbers) • ISSN 0065-9266

American Mathematical Society
Providence, Rhode Island

2000 *Mathematics Subject Classification.* Primary 16W30; Secondary 17B37.

Library of Congress Cataloging-in-Publication Data

Natale, Sonia, 1972–
    Semisolvability of semisimple Hopf algebras of low dimension / Sonia Natale.
      p. cm. — (Memoirs of the American Mathematical Society, ISSN 0065-9266 ; no. 874)
    "Volume 186, number 874 (fourth of 5 numbers)."
    Includes bibliographical references.
    ISBN-13: 978-0-8218-3948-5 (alk. paper)
    ISBN-10: 0-8218-3948-9 (alk. paper)
    1. Hopf algebras.   2. Quantum groups.   I. Title.

QA613.8.N38   2007
512′.55—dc22
                                                                                                    2006047928

Memoirs of the American Mathematical Society

This journal is devoted entirely to research in pure and applied mathematics.

Subscription information. The 2007 subscription begins with volume 185 and consists of six mailings, each containing one or more numbers. Subscription prices for 2007 are US$649 list, US$519 institutional member. A late charge of 10% of the subscription price will be imposed on orders received from nonmembers after January 1 of the subscription year. Subscribers outside the United States and India must pay a postage surcharge of US$38; subscribers in India must pay a postage surcharge of US$43. Expedited delivery to destinations in North America US$53; elsewhere US$130. Each number may be ordered separately; *please specify number* when ordering an individual number. For prices and titles of recently released numbers, see the New Publications sections of the *Notices of the American Mathematical Society*.

Back number information. For back issues see the *AMS Catalog of Publications*.

Subscriptions and orders should be addressed to the American Mathematical Society, P. O. Box 845904, Boston, MA 02284-5904, USA. *All orders must be accompanied by payment.* Other correspondence should be addressed to 201 Charles Street, Providence, RI 02904-2294, USA.

Copying and reprinting. Individual readers of this publication, and nonprofit libraries acting for them, are permitted to make fair use of the material, such as to copy a chapter for use in teaching or research. Permission is granted to quote brief passages from this publication in reviews, provided the customary acknowledgment of the source is given.

Republication, systematic copying, or multiple reproduction of any material in this publication is permitted only under license from the American Mathematical Society. Requests for such permission should be addressed to the Acquisitions Department, American Mathematical Society, 201 Charles Street, Providence, Rhode Island 02904-2294, USA. Requests can also be made by e-mail to reprint-permission@ams.org.

Memoirs of the American Mathematical Society is published bimonthly (each volume consisting usually of more than one number) by the American Mathematical Society at 201 Charles Street, Providence, RI 02904-2294, USA. Periodicals postage paid at Providence, RI. Postmaster: Send address changes to Memoirs, American Mathematical Society, 201 Charles Street, Providence, RI 02904-2294, USA.

© 2007 by the American Mathematical Society. All rights reserved.
This publication is indexed in *Science Citation Index*®, *SciSearch*®, *Research Alert*®, *CompuMath Citation Index*®, *Current Contents*®/*Physical, Chemical & Earth Sciences*.
Printed in the United States of America.

∞ The paper used in this book is acid-free and falls within the guidelines established to ensure permanence and durability.
Visit the AMS home page at http://www.ams.org/

10 9 8 7 6 5 4 3 2 1   12 11 10 09 08 07

Contents

Introduction and Main Results	1
Conventions and Notation.	7
Chapter 1. Semisimple Hopf Algebras	9
1.1. Algebra structure	9
1.2. Irreducible characters	10
1.3. Coinvariants of Hopf algebra maps	12
1.4. Yetter-Drinfeld modules	13
1.5. Yetter-Drinfeld modules and the character algebra	14
1.6. One dimensional Yetter-Drinfeld modules	15
1.7. $H^{\operatorname{co} B}$ as a left coideal of H	16
Chapter 2. The Nichols-Richmond Theorem	17
2.1. Irreducible characters of degree 2	17
2.2. The Nichols-Richmond theorem	19
2.3. An application to $D(H)$	20
2.4. Existence of proper Hopf subalgebras	20
2.5. Hopf subalgebras of index 3	22
Chapter 3. Quotient Coalgebras	25
3.1. A multiplicity formula	25
3.2. Stable subcoalgebras	27
3.3. Quotients modulo group-like Hopf subalgebras	30
3.4. On the structure of $G(H)$	31
3.5. A criterion of normality	32
Chapter 4. Braided Hopf Algebras	33
4.1. Radford-Majid biproduct construction	33
4.2. Coalgebra structure of R	34
4.3. Hopf subalgebras	36
4.4. Biproducts over finite groups	37
4.5. Cocommutative braided Hopf algebras	39
4.6. Cocommutative braided Hopf algebras over $\mathbb{Z}_p$	40
4.7. Transitive actions of central subgroups	43

Chapter 5. Cocycle Deformations of Some Hopf Algebras	45
5.1. Lifting from abelian groups	45
5.2. Examples in dimension pq^2; $p \equiv 1 \mod q$	46
5.3. Examples in dimension pq^2; $q \equiv 1 \mod p$	47
5.4. Normal Hopf subalgebras in cocycle twists	47
Chapter 6. Dimension 24	49
6.1. Possible (co)-algebra structures	49
6.2. Upper and lower semisolvability	54
Chapter 7. Dimension 30	57
7.1. Possible (co)-algebra structures	57
7.2. Classification	60
Chapter 8. Dimension 36	61
8.1. Reduction of the problem	61
8.2. Main result	67
Chapter 9. Dimension 40	69
9.1. Reduction of the problem	69
9.2. Main result	73
Chapter 10. Dimension 42	75
10.1. Possible (co)-algebra structures	75
10.2. Classification	79
Chapter 11. Dimension 48	81
11.1. First reduction	81
11.2. Further reductions	92
11.3. Main result up to cocycle twists	96
11.4. Main result	101
Chapter 12. Dimension 54	107
12.1. First reduction	107
12.2. Main result	111
Chapter 13. Dimension 56	113
13.1. First reduction	113
13.2. Main result	115
Appendix A. Drinfeld Double of H_8	117
A.1. Structure of $D(H_8)$	117
A.2. Proof of Theorem A.1.1	118
A.3. Proof of Theorem A.1.2	120
Appendix. Bibliography	121

Abstract

We prove that every semisimple Hopf algebra of dimension less than 60 over an algebraically closed field k of characteristic zero is either upper or lower semisolvable up to a cocycle twist.

Received by the editor September 2006.
1991 *Mathematics Subject Classification.* Primary 16W30; Secondary 17B37.
Key words and phrases. semisimple Hopf algebra; semisolvability; Hopf algebra extension.
This work was partially supported by CONICET, Agencia Córdoba Ciencia, ANPCyT, Fundación Antorchas and Secyt (UNC).

Introduction and Main Results

In recent papers several notions and results from the theory of finite groups have been generalized or adapted to the context of (semisimple) Hopf algebras. Simultaneously, many results on the classification of semisimple Hopf algebras have also appeared. Some conjectural analogies, such as Kaplansky's conjecture about the dimensions of the irreducible modules, still remain an open problem.

Let H be a finite dimensional Hopf algebra over a field k. A Hopf subalgebra $A \subseteq H$ is called *normal* if $h_1 A \mathcal{S}(h_2) \subseteq A$, for all $h \in H$. If H does not contain proper normal Hopf subalgebras then it is called *simple*.

If $A \subseteq H$ is a normal Hopf subalgebra then the structure of H can be reconstructed from A and the quotient Hopf algebra $\overline{H} = H/HA^+$; more precisely, it is known that in this case H is isomorphic to a bicrossed product $H \simeq A \#_{\rightharpoonup, \sigma}^{\rho, \tau} \overline{H}$, where $(\rightharpoonup, \sigma, \rho, \tau)$ is a *compatible datum*; see for instance [**A, M, M10**]. This fact implies that, when trying to classify Hopf algebras of a given finite dimension, it is an important problem to decide whether the Hopf algebra is simple or not.

We shall assume from now on that the field k is algebraically closed of characteristic zero.

We say that a finite dimensional Hopf algebra H is *trivial* if it is isomorphic to a group algebra or to a dual group algebra. Then, H is trivial if and only if it is commutative or cocommutative.

The notions of upper and lower semisolvability for finite-dimensional Hopf algebras have been introduced in [**MW**], as generalizations of the notion of solvability for finite groups. By definition, H is called *lower semisolvable* if there exists a chain of Hopf subalgebras $H_{n+1} = k \subseteq H_n \subseteq \cdots \subseteq H_1 = H$ such that H_{i+1} is a normal Hopf subalgebra of H_i, for all i, and all *factors* $\overline{H}_i := H_{i+1}/H_{i+1}H_i^+$ are trivial. Dually, H is called *upper semisolvable* if there exists a chain of quotient Hopf algebras $H_{(0)} = H \to H_{(1)} \to \cdots \to H_{(n)} = k$ such that each of the maps $H_{(i-1)} \to H_{(i)}$ is normal, and all *factors* $H_i := H_{(i-1)}^{\text{co}\pi_i}$ are trivial; here, $H_{(i-1)}^{\text{co}\pi_i}$ is the space of coinvariants of the map π_i, see Section 1.3.

We have that H is upper semisolvable if and only if H^* is lower semisolvable [**MW**]. If this is the case, then H can be obtained from group algebras and their duals by means of (a finite number of) extensions; in particular, H is semisimple.

The smallest non-solvable group is the simple alternating group $\mathbb{A}_5$ of order 60. It is thus natural to ask if an analogous statement is true for semisimple Hopf algebras. A version of the following question was posed by S. Montgomery in [**Mo1**, Question, pp. 269].

QUESTION 1. *Let H be a semisimple Hopf algebra of dimension less than* 60. *Is H necessarily upper or lower semisolvable?*

Let H be a semisimple Hopf algebra over k. If $\dim H = p^n$, where p is a prime number, then H has a nontrivial central group-like element [**M6**]; inductively, one can see that H is both upper and lower semisolvable [**MW**]. Also, if $\dim H = pq^2$, where $p \neq q$ are prime numbers, then it was shown in [**N, N2, N3**] that, under the assumption that H and H^* are both of Frobenius type, either H or H^* contains a nontrivial central group-like element. This implies that these Hopf algebras are also semisolvable, since semisimple Hopf algebras of dimension p, pq and q^2 are trivial. In [**N3**] we showed that all semisimple Hopf algebras of dimension $pq^2 < 100$ are of Frobenius type (some instances of this fact, e.g., dimension 44, appeared in [**K2**]); so that these are all semisolvable.

However, not every nontrivial semisimple Hopf algebra H is semisolvable. An example of a simple nontrivial semisimple Hopf algebra H of dimension 60 and was constructed by D. Nikshych in [**Nk**]: in this case H is a cocycle twist of the group algebra of the simple group $\mathbb{A}_5$. Moreover, it was shown in [**Nk**] that if G is a finite simple group and $\phi \in kG \otimes kG$ is a nontrivial invertible pseudo 2-cocycle, then the twisted group algebra $(kG)_\phi$ is a nontrivial semisimple Hopf algebra, which is simple as a Hopf algebra.

The smallest example of a semisimple Hopf algebra which is not semisolvable is a cocycle twist of a group of order 36 [**GN**]. So the answer to Question 1 is negative, and it can only be expected to be affirmative up to a cocycle twist. The dimensions where the problem remains open are 24, 30, 36, 40, 42, 48, 54 and 56. We refer the reader to [**A2, Mo1**] for an account of previous results on the problem of classification.

We also point out that, in the related context of Kac algebras, several classification results in low dimension were obtained by Izumi and Kosaki in their work [**IK**]; in that paper, the authors classify all Kac algebras of dimensions 16, 24, $pq^2 < 60$ and $pqr < 60$.

Our main result is the following theorem, giving an affirmative answer to Question 1 up to cocycle twists.

THEOREM 1. *Let H be a semisimple Hopf algebra of dimension < 60. Then H is either upper or lower semisolvable up to a cocycle twist.*

We prove that a semisimple Hopf algebra of dimension 24, 30, 40, 42, 48, 54 or 56 is *not simple*, and moreover, in dimension 36 the only simple example is a twisting of a finite group. This is equivalent to the statement in the theorem, in view of previous results. Indeed, if H is a nontrivial semisimple Hopf algebra of dimension < 60, then the following are equivalent (see [**N4**]):

- H is not simple;
- H is either upper or lower semisolvable.

Our approach to the problem is the following: for each fixed dimension, we first consider the possible algebra and coalgebra decompositions (which turn out to be of Frobenius type). Then, for each possible type, we derive the existence of proper normal Hopf subalgebras.

In order to do this, we discuss some general result on semisimple Hopf algebras. Many of these results are new. We discuss some properties of irreducible characters of low degree which allow, in most cases, to prove the existence of quotient Hopf algebras or Hopf subalgebras, for each fixed algebra or coalgebra structure, respectively. One of the main tools for this is the use of the Nichols-Richmond theorem on irreducible characters of degree 2 [**NR**] and some of its consequences, which we develope in Chapter 2.

Using the character theory of H, we get a general result on Hopf subalgebras with index 3, which is often of use in low dimensions. See Theorem 2.5.1.

In Chapter 3 we consider an inclusion $A \subseteq H$ of semisimple Hopf algebras. We show that if C is a simple subcoalgebra of H such that $Ca \subseteq C$, for all $a \in A$, then the dual of the quotient coalgebra C/CA^+ and the crossed product A_α, where $\alpha : A \otimes A \to k$ is a certain 2-cocycle, constitute a commuting pair in C^*.

This relates the corepresentation theory of C/CA^+ with the representation theory of the Galois object A_α of A. Applied in combination with Masuoka's main result in [**M5**] on deformations of cosemisimple Hopf algebras, the result allows us to prove that some coalgebra decompositions are impossible. The result is also useful to get some information on the structure of the Hopf subalgebra A, especially when A is a group algebra; see Section 3.4.

We also discuss braided Hopf algebras in relation with the Radford-Majid biproduct construction; see Chapter 4. It often happens, mainly because of the 'self-dual' nature of the assumption of simplicity, that if a given semisimple Hopf algebra H is simple, then H must have the structure of a biproduct $H = R \# A$, where A is a semisimple Hopf subalgebra. We get several results on existence of proper (normal) Hopf subalgebras in biproducts; in particular, we give such a result in Corollary 4.3.4 for the case where A is not cocommutative of dimension p^3, p a prime number.

Let $p \neq q$ be prime numbers. We describe the known families in dimension pq^2 as cocycle twists of group algebras; we also prove that other families cannot be obtained in this fashion. See Chapter 5.

The classification of semisimple Hopf algebras of dimension pqr, where p, q and r are distinct prime numbers, was given in [**N**] under the assumption that H admits an extension with commutative 'kernel' and cocommutative 'cokernel' (a so called *abelian* extension).

In this paper we also prove that semisimple Hopf algebras of dimension 30 and 42 admit abelian extensions. This allows us to give the complete classification of semisimple Hopf algebras of these dimensions. We obtain the following theorems. See Chapters 7, 10.

THEOREM 2. *Let H be a semisimple Hopf algebra of dimension* 30 *over k. Then H is isomorphic to a group algebra kG or to a dual group algebra k^G, where G is a group of order* 30.

The known nontrivial examples in dimension 42 were constructed in [**AN**]: these are denoted $\mathcal{A}_7(2,3)$ and $\mathcal{A}_7(3,2) \simeq \mathcal{A}_7(2,3)^*$. We prove in this paper the following result.

THEOREM 3. *Let H be a nontrivial semisimple Hopf algebra of dimension* 42 *over k. Then H is isomorphic to one of the Hopf algebras* $\mathcal{A}_7(2,3)$ *or* $\mathcal{A}_7(3,2)$.

In Table 1 we resume some known facts about the classification of semisimple Hopf algebras of dimension less than 60. In the first column, p, q and r are distinct prime numbers. The table is organized as follows: the first column indicates the factorizations of the dimensions ≤ 60 in terms of prime numbers; the second and third column contain the references where nontrivial examples were constructed and where the classification was given, respectively. The fourth column gives additional information for each specific type.

We include an appendix where we describe the structure of the Drinfeld doubles of the three non-commutative semisimple Hopf algebras of dimension 8. Tambara and Yamagami show in [**TY**] that the

dim H	Nontrivial examples	Classification	Remarks
p	–	[**Ka2, Z**]. See also [**ZS**].	$H \simeq k\mathbb{Z}_p$.
p^2	–	[**M7**].	$H \simeq kG$, G abelian of order p^2
pq	–	[**M4, EG, GW**].	$H \simeq kG$, or k^G. Other proofs in [**So, N**].
p^3	[**KP**] for dim 8. [**M7**] for dim 27.	[**M4, M7**].	All semisolvable by [**M6, MW**].
pq^2	[**F, G, M3, N**].	[**F**] for dim 12, [**M3**] for dim 18, [**N, N2, N3**].	All semisolvable. [**IK**, Chapter X] for Kac algebras.
p^4 (16)	[**K**].	[**K**].	All semisolvable by [**M6, MW**].
p^3q (24, 40, 54, 56)	[**IK**] for dim 24.	[**IK**, Chapter XIV] for Kac algebras of dim 24.	Semisolvable; Chapters 6, 9, 12, 13.
pqr (30, 42)	[**AN**] for dim 42. No nontrivial example in dim 30.	Chapters 7, 10.	Abelian extensions in dim pqr classified in [**N**]. [**IK**, Chapter X] for Kac algebras.
p^5 (32)	Yes.	–	All semisolvable by [**M6, MW**].
p^2q^2 (36)	[**EG4**]; $D(\mathbb{S}_3)$; $(kD_3 \times D_3)_\phi$ simple [**GN**]; more.	–	First non-semisolvable example. Chapter 8.
p^4q (48)	Yes.	–	Semisolvable; Chapter 11
p^2qr (60)	$(k\mathbb{A}_5)_\phi$ (simple) [**Nk**]; $(kD_3 \times D_5)_\phi$ (simple) [**GN**].	–	–

TABLE 1. Semisimple Hopf algebras of dimension ≤ 60

categories of representations of these three Hopf algebras are not equivalent as monoidal categories. The results in this appendix have been motivated by the paper [**Mo3**], where the Schur indicators for the three Hopf algebras are compared: this gives evidence that the representation theory of H_8 is in some sense closer to that of kD_4 than to that of kQ. Here, H_8 denotes the unique nontrivial 8-dimensional semisimple Hopf algebra over k [**KP, M4**], while D_4 and Q denote, respectively, the dihedral and quaternionic groups of order 8.

On the other hand, note that H_8 is an extension of $k^{\mathbb{Z}_2 \times \mathbb{Z}_2}$ by $k\mathbb{Z}_2$, such that the bicrossproduct group $(\mathbb{Z}_2 \times \mathbb{Z}_2) \bowtie \mathbb{Z}_2$ associated to the matched pair of the extension is D_4. As a consequence of [**N5**, Theorem 1.3], we know that $D(H_8)$ is a cocycle twist of the Dijkgraaf-Pasquier-Roche quasi-Hopf algebra $D^\omega(D_4)$, where $\omega \in H^3(D_4, k^\times)$ is the 3-cocycle associated to the extension corresponding to H_8 via the Kac exact sequence [**Ka**]. We present more evidence of this facts involving the Drinfeld doubles. More precisely, we prove on the one hand that $D(H_8)$ has no quotient Hopf algebra isomorphic to kQ. We also show that $D(H_8)$ is a (central) extension of k^G by kG, where $G = G(D(H_8)^*) \simeq \mathbb{Z}_2 \times \mathbb{Z}_2 \times \mathbb{Z}_2$. See Theorems A.1.1, A.1.2.

Acknowledgements. Most of the results in this paper have been announced in [**N4**]. They were also communicated in the conferences *Hopf algebras in Noncommutative Geometry and Physics* (Brussels, May 2002) and *Colloquium on Homology Theories, Representations and Hopf Algebras* (Luminy, June 2002).

The author is grateful to N. Andruskiewitsch, S. Montgomery, L. Vainerman and H.-J. Schneider for interesting discussions, comments and references. She also thanks Y. Kashina and Y. Sommerhäuser for helpful remarks on a previous version of this paper, and the referee for many valuable comments.

This research has been done during a postdoctoral stay at the Department of Mathematics of the École Normale Supérieure, Paris. The author is grateful to Marc Rosso for his kind hospitality.

Conventions and Notation.

Throughout, k will denote an algebraically closed field of characteristic zero. The symbols Hom, $\otimes$, etc., will mean Hom_k, $\otimes_k$, etc. Our references for the theory of Hopf algebras are [**Mo, Sc**]. The notation for Hopf algebras is standard; for instance, the group of group-like elements in H is denoted by $G(H)$. For an algebra A (respectively, for a coalgebra C) the notation Hom_A (resp. Hom^C) is used to indicate the Hom bifunctor in the category of (left) A-modules (resp. C-comodules).

A Hopf algebra H is called *semisimple* (respectively, *cosemisimple*) if it is semisimple as an algebra (respectively, if it is cosemisimple as a coalgebra). Let H be a finite-dimensional Hopf algebra over k. By a result of Larson and Radford, it is known that H is semisimple if and only if H is cosemisimple, if and only if $\mathcal{S}^2 = \text{id}$. See [**LR, LR2**]. The *character algebra* of H, denoted $R(H)$, is the subalgebra of H^* spanned by the irreducible characters of H; if H is semisimple, $R(H)$ coincides with the subalgebra of cocommutative elements in H^*.

Suppose H is finite dimensional. For a Hopf subalgebra $A \subseteq H$, the *index* of A in H is defined by $[H : A] := \dfrac{\dim H}{\dim A}$; it is an integer by [**NZ**]. Suppose $q : H \to B$ is a surjective Hopf algebra map, and identify B^* with a Hopf subalgebra of H^* via q^*; by abuse of terminology, the index $[H^* : B^*]$ will be also called the index of B in H.

CHAPTER 1

Semisimple Hopf Algebras

Along this chapter, H will be a semisimple (thus finite-dimensional) Hopf algebra over k.

1.1. Algebra structure

As an algebra, H is isomorphic to a direct product of full matrix algebras

$$(1.1.1) \qquad H \simeq k^{(n)} \times \prod_{d_i > 1} M_{d_i}(k)^{(n_i)},$$

where $n = |G(H^*)|$. It follows from the Nichols-Zoeller Theorem [**NZ**], that n divides both $\dim H$ and $n_i d_i^2$, for all i. Moreover, by [**NR**] if $d_i = 2$ for some i, then the dimension of H is even.

By [**ZS**], if $n = 1$, then $\{d_i : d_i > 1\}$ has at least three elements.

Suppose that $A \subseteq H$ is a Hopf subalgebra. Then A is also semisimple. Assume that A is commutative. Then it follows from the Frobenius Reciprocity that $d_i \leq [H : A]$, for all i. See [**AN2**, Corollary 3.9].

If H is as in (1.1.1) as an algebra, we shall say that H is *of type* $(1, n; d_1, n_1; \ldots; d_r, n_r)$ *as an algebra*. In this case, the dimension of the character algebra of H is $n + n_1 + \cdots + n_r$.

If H^* is of type $(1, n; d_1, n_1; \ldots)$ as an algebra, we shall say that H is *of type* $(1, n; d_1, n_1; \ldots)$ *as a coalgebra*.

So that H is of type $(1, n; d_1, n_1; \ldots; d_r, n_r)$ as a (co-)algebra if and only if H has n non-isomorphic one-dimensional (co-)representations, n_1 non-isomorphic irreducible (co-)representations of degree d_1, etc. Sometimes, we shall use the notation X_d or $X_d(H)$ to indicate the set of irreducible characters of H of degree d.

EXAMPLE 1.1.2. The above arguments can be used to get the possible algebra structures for a given finite dimension. For instance, suppose that B is a semisimple Hopf algebra of dimension 60 such that $G(B^*) = 1$. Then, as an algebra, B is of one of the following types:

$(1, 1; 3, 2; 4, 1; 5, 1)$, $(1, 1; 2, 4; 3, 2; 5, 1)$ or $(1, 1; 2, 4; 3, 3; 4, 1)$.

1.2. Irreducible characters

Let V be an H-module. The *character* of V is the element $\chi = \chi_V \in H^*$ defined by $\langle \chi, h \rangle = \operatorname{Tr}_V(h)$, $h \in H$. The *degree* of χ is the integer $\deg \chi = \chi(1) = \dim V$. If U is another H-module, we have

$$\chi_{V \otimes U} = \chi_V \chi_U, \qquad \chi_{V^*} = \mathcal{S}(\chi_V).$$

Thus the irreducible characters, *i.e.*, the characters of the irreducible H-modules, span a subalgebra $R(H)$ of H^*, called the *character algebra* of H. The antipode induces an anti-algebra involution $* : R(H) \to R(H)$, $\chi \mapsto \chi^* := \mathcal{S}(\chi)$. The degree defines an augmentation $R(H) \to k$.

We first resume some of the basic properties of $R(H)$ that will be often used in the rest of this paper. Proofs of these facts can be found in [**NR**].

Let $\chi_V, \chi_W \in R(H)$ be the characters of the H-modules V and W, respectively. The integer $m(\chi_V, \chi_W) = \dim \operatorname{Hom}_H(V, W)$ will be called *multiplicity* of V in W. This extends to a bilinear form $m : R(H) \times R(H) \to k$.

Let $\widehat{H}$ denote the set of irreducible characters of H. If $\chi \in R(H)$, then we may write $\chi = \sum_{\mu \in \widehat{H}} m(\mu, \chi) \mu$. Let χ, ψ and λ be characters of H-modules; we have

(1.2.1) $$m(\chi, \psi \lambda) = m(\psi^*, \lambda \chi^*) = m(\psi, \chi \lambda^*).$$

Let χ be an irreducible character of H. Denote by $G[\chi]$ the subgroup of $G(H^*)$ consisting of all those elements g such that $g\chi = \chi$. We have

(1.2.2) $$g \in G[\chi] \iff m(g, \chi \chi^*) > 0 \iff m(g, \chi \chi^*) = 1.$$

See [**NR**, Theorem 10]. In particular,

(1.2.3) $$\chi \chi^* = \sum_{g \in G[\chi]} g + \sum_{\mu \in \widehat{H}, \deg \mu > 1} m(\mu, \chi \chi^*) \mu.$$

Note that $G[\chi^*] = \{g \in G(H^*) : \chi g = \chi\}$. Also, if $a \in G(H^*)$, we have

$$G[\chi a] = G[\chi], \qquad G[a\chi] = a G[\chi] a^{-1}.$$

As a consequence of [**NZ**], we have that $|G[\chi]|/(\deg \chi)^2$. Moreover, it follows from the results in [**M5**, Section 2], that the order of any element $g \in G[\chi]$ (hence also the exponent of $G[\chi]$) divides $\deg \chi$.

The following lemma will be applied later; see Lemma 12.1.1. It serves here to illustrate the (mostly well-known) fact that not any algebra type can arise as the structure of a semisimple Hopf algebra.

LEMMA 1.2.4. *There is no semisimple Hopf algebra with algebra type* $(1, 2; 2, 1; 4, m)$, $m \geq 1$.

PROOF. Suppose on the contrary that H is a semisimple Hopf algebra with this algebra type. Let $G(H^*) = \{\epsilon, g\}$ and let χ be the unique irreducible character of degree 2 of H; so that we have $g\chi = \chi = \chi g$, $\chi^* = \chi$ and $\chi^2 = \epsilon + g + \chi$. In particular, $m(\chi, \zeta\chi) = m(\zeta, \chi^2) = 0$, for all irreducible character ζ of degree 4. Then also $m(\chi, \chi\zeta) = m(\chi, \zeta^*\chi) = 0$, for all such ζ.

Let $\deg \zeta = 4$ and write $\zeta\zeta^* = \epsilon + g + n\chi + \lambda$, where λ is a character of H such that $m(\chi, \lambda) = 0$. By taking degrees, we get that $n > 0$ and n is odd. Since $n = m(\zeta, \chi\zeta)$, it follows that $n = 1$. Therefore, $\chi\zeta = \zeta + \psi$, where $\deg \psi = 4$, $\psi \neq \zeta$. Since $m(\chi, \chi\zeta) = 0$, then ψ is irreducible.

Then $m(\epsilon, \chi\zeta\psi^*) = m(\epsilon, \psi\psi^*) = 1$, and we have $m(\chi, \zeta\psi^*) = 1$. Thus $\zeta\psi^* = \chi + \sum_l \psi_l$, where $\deg \psi_l = 4$. Taking degrees we get a contradiction. This shows that this type is not possible. □

Let $n \geq 1$. The group $G(H^*) \times G(H^*)$ acts on the set X_n via

(1.2.5) $\qquad (g, h).\chi = g\chi h^{-1}, \qquad g, h \in G(H^*), \quad \chi \in X_p.$

We have $G[(g, h).\chi] = gG[\chi]g^{-1}$, for all $g, h \in G(H^*)$, and for all χ.

Using this action we can get some information on the structure of the group $G(H^*)$. The following proposition gives an example of this fact; see also the proof of Lemma 10.1.6.

PROPOSITION 1.2.6. *Let $p < q$ be prime numbers. Suppose that $G(H^*)$ is nonabelian of order pq. Assume in addition that $G[\chi] \neq 1$, for all $\chi \in X_p$. Then q^2 divides $|X_p|$.*

PROOF. We may assume that $X_p \neq \emptyset$. Let $T \subseteq G(H^*)$ be a subgroup of order q. Consider the action $(T \times T) \times X_p \to X_p$ obtained by restriction of the action (1.2.5). We shall show that the stabilizer $(T \times T)_\chi$ is trivial, for all $\chi \in X_p$, which will imply the proposition.

Let $g, h \in T$, $\chi \in X_p$, and suppose that $(g, h).\chi = g\chi h^{-1} = \chi$. Then $G[\chi] = G[(g, h).\chi] = gG[\chi]g^{-1}$. By assumption, $G[\chi] \neq 1$, hence $|G[\chi]| = p$, because of the assumption $|G(H^*)| = pq$. Then $g = 1$; otherwise, $G[\chi]$ would be a normal subgroup of order p in $G(H^*)$, implying that $G(H^*)$ is abelian against the assumptions.

Hence we have $(g, h).\chi = \chi h^{-1} = \chi$, and $h^{-1} \in G[\chi^*]$. Thus also $h = 1$. This finishes the proof of the proposition. □

1.3. Coinvariants of Hopf algebra maps

Let $q : H \to B$ be a Hopf algebra map and consider the subspaces of coinvariants

$$H^{co\, q} = \{h \in H : (\mathrm{id} \otimes q)\Delta(h) = h \otimes 1\}, \quad \text{and}$$
$$^{co\, q}H = \{h \in H : (q \otimes \mathrm{id})\Delta(h) = 1 \otimes h\}.$$

Then $H^{co\, q}$ (respectively, $^{co\, q}H$) is a left (respectively, right) coideal subalgebra of H. We shall also use the notation $H^{co\, B} := H^{co\, q}$. By [**Sc2**],

(1.3.1) $\qquad \dim H = \dim H^{co\, q} \dim q(H) = \dim {}^{co\, q}H \dim q(H).$

The left coideal subalgebra $H^{co\, q}$ is stable under the left adjoint action of H. Moreover $H^{co\, q} = {}^{co\, q}H$ if and only if $H^{co\, q}$ is a (normal) Hopf subalgebra of H. If this is the case, we shall say that the map $q : H \to B$ is normal.

REMARK 1.3.2. The Hopf algebra B^* acts on H on the left and on the right by $f \rightharpoonup h = \langle f, q(h_2) \rangle h_1$, and $h \leftharpoonup f = \langle f, q(h_1) \rangle h_2$, respectively. Suppose that q is surjective. Then we have

$$H^{co\, q} = \{h \in H : f \rightharpoonup h = \epsilon(f)h, \forall f \in B^*\}, \quad \text{and}$$
$$^{co\, q}H = \{h \in H : h \leftharpoonup f = \epsilon(f)h, \forall f \in B^*\}.$$

In particular, let $\eta \in G(H^*)$ and consider the Hopf algebra map $q : H \to k^{\langle \eta \rangle}$ obtained by transposing the inclusion $k\langle \eta \rangle \subseteq H^*$. Then

$$H^{co\, q} = \{h \in H : \eta \rightharpoonup h = h\}, \qquad {}^{co\, q}H = \{h \in H : h \leftharpoonup \eta = h\},$$

where $\leftharpoonup$ and $\rightharpoonup$ are the regular actions of H^* on H.

By [**NZ**], if A is a Hopf subalgebra of H such that $A \subseteq H^{co\, q}$, then $H^{co\, q}$ is free as an A-module, with respect to the action by left multiplication of A. In particular, $\dim A$ divides $\dim H^{co\, q}$. The same holds true with $^{co\, q}H$ instead of $H^{co\, q}$. Indeed, with this A-module structure and the coaction given by the comultiplication of H, $H^{co\, q}$ is a left (A, H) Hopf module. We note the following consequence of this fact:

LEMMA 1.3.3. *Suppose that $|G(H)|$ and $[H^* : G(H^*)]$ are relatively prime. Then the group $G(H)$ is abelian and isomorphic to a subgroup of $\widehat{G(H^*)}$.*

PROOF. There is a surjective Hopf algebra map $\pi : H \to k^{G(H^*)}$, and we have $\dim H^{co\, \pi} = [H^* : G(H^*)]$. If $1 \neq g \in G(H)$, then $\pi(g) \neq 1$, since otherwise g would belong to $H^{co\, \pi}$ implying that the order of g divides $\dim H^{co\,\pi}$, contradicting the assumption. Therefore the restriction of π to $G(H)$ is injective, and $G(H)$ is thus isomorphic to a subgroup of $\widehat{G(H^*)} = G(k^{G(H^*)})$. This implies that $G(H)$ is abelian. □

LEMMA 1.3.4. *Suppose that $A \subseteq H$ is a Hopf subalgebra. Then $A^{co\,q|_A} = A \cap H^{co\,q}$. In particular, $\dim A = \dim(A \cap H^{co\,q})\,\dim q(A)$.*

PROOF. The first claim is evident. The second follows from (1.3.1). □

1.4. Yetter-Drinfeld modules

Let ${}^H_H\mathcal{YD}$ denote the category of (left-left) Yetter-Drinfeld modules over H. Objects of this category are vector spaces V endowed with an H-coaction $\rho : V \to H \otimes V$ and an H-action $.: H \otimes V \to V$, subject to the compatibility condition $\rho(h.v) = h_1 v_{-1} \mathcal{S}(h_3) \otimes h_2.v_0$, $v \in V$, $h \in H$; morphisms are H-linear and colinear maps.

The category ${}^H_H\mathcal{YD}$ is a modular category, which coincides as such with the category of modules over the Drinfeld double of H; see [**EG**]. It is shown in [**EG**] that if V is a simple Yetter-Drinfeld module over H, then $\dim V$ divides $\dim H$.

With respect to the left adjoint action $\mathrm{ad} : H \otimes H \to H$, $(\mathrm{ad}\,h)(a) = h_1 a \mathcal{S}(h_2)$ and the left regular coaction $\Delta : H \to H \otimes H$, H becomes an object of ${}^H_H\mathcal{YD}$.

The Yetter-Drinfeld submodules $V \subseteq H$ are exactly the left coideals V of H such that $h_1 V \mathcal{S}(h_2) \subseteq V$, for all $h \in H$. Thus, a one-dimensional Yetter-Drinfeld submodule of H is exactly the span of a central group-like element of H.

It is well-kown that the space of (right) coinvariants of a Hopf algebra map is a left coideal stable under the left adjoint action. We thus obtain the following lemma:

LEMMA 1.4.1. *Let $H \to B$ be a Hopf algebra map. Then $H^{co\,B}$ is a Yetter-Drinfeld submodule of H.* □

REMARK 1.4.2. Suppose that $H \to B \to B'$ is a sequence of Hopf algebra maps. Then $H^{co\,B}$ is a Yetter-Drinfeld submodule of $H^{co\,B'}$. In particular, since ${}^H_H\mathcal{YD}$ is a semisimple category, there exists a Yetter-Drinfeld submodule $W \subseteq H^{co\,B'}$ such that $H^{co\,B'} = H^{co\,B} \oplus W$.

We recover the following result, due to Kobayashi and Masuoka [**KM**]; see also [**N**, Theorem 2.1.1]. Our alternative proof is based on [**EG**].

COROLLARY 1.4.3. *Suppose that $B \subseteq H$ is a Hopf subalgebra such that $[H : B] = p$ is the smallest prime number dividing $\dim H$. Then B is a normal Hopf subalgebra and H fits into a (co-)central extension $1 \to B \to H \to k\mathbb{Z}_p \to 0$.* □

Our argument proves indeed the following more precise statement: if $A \subseteq H$ is a *normal* Hopf subalgebra such that $\dim A$ is the smallest prime number dividing $\dim H$, then A is *central* in H.

PROOF. Consider the dual projection $H^* \to B^*$. So that we have $\dim(H^*)^{\operatorname{co} B^*} = [H : B] = p$. Let V be an irreducible Yetter-Drinfeld submodule of $(H^*)^{\operatorname{co} B^*}$. Since the dimension of V divides $\dim H$ and is less than p, we find that $\dim V = 1$; therefore $V = kg$, for some $g \in Z(H^*) \cap G(H^*)$.

Decomposing $(H^*)^{\operatorname{co} B^*}$ into irreducible Yetter-Drinfeld modules, we see that $(H^*)^{\operatorname{co} B^*}$ is a central group-like Hopf subalgebra of H^* of dimension p. This implies that H^* fits into a central extension $0 \to k\mathbb{Z}_p \to H^* \to B^* \to 1$. The lemma follows after dualizing this extension. □

1.5. Yetter-Drinfeld modules and the character algebra

In the paper [**Z2**] Y. Zhu establishes a bijective correspondence between primitive idempotents in the character algebra $R(H) \subseteq H^*$ and irreducible Yetter-Drinfeld submodules of H. Indeed, it is shown in [**Z2**] that $D(H)$ and $R(H)$ form a commuting pair in $\operatorname{End} H$, with respect to the $D(H)$-action corresponding to the Yetter-Drinfeld module structure in H considered in Section 1.4 (or a version of it thereof) and the $R(H)$-action $\rightharpoonup: R(H) \otimes H \to H$, $f \rightharpoonup h := \langle f, h_2 \rangle h_1$.

Let $q : H \to B$ be a Hopf algebra projection. Then $H^{\operatorname{co} B}$ is a Yetter-Drinfeld submodule of H. Consider the dual inclusion of Hopf algebras $B^* \to H^*$ and let $e_0 \in B^*$ be the normalized integral; e_0 is the primitive idempotent in B^* corresponding to the trivial representation. Since e_0 is a cocommutative element, we have $e_0 \in R(H)$. Hence we may write

(1.5.1) $$e_0 = \Lambda + E_1 + \cdots + E_n,$$

where $\Lambda, E_1, \ldots, E_n$ are orthogonal primitive idempotents in $R(H)$, such that Λ is the normalized integral in H^*.

The following proposition gives a refinement of the result in [**Z2**].

PROPOSITION 1.5.2. *The idempotents $\Lambda, E_1, \ldots, E_n$ in (1.5.1) correspond bijectively with the irreducible H-Yetter-Drinfeld submodules of $H^{\operatorname{co} B}$.*

1.6. One dimensional Yetter-Drinfeld modules

PROOF. We saw in Remark 1.3.2 that $H^{\operatorname{co} B}$ coincides with the subalgebra of B^*-invariant elements of H under the left regular action $\rightharpoonup : B^* \otimes H \to H$. Hence, since $e_0 \in B^*$ is the primitive idempotent corresponding to the trivial representation, we have $H^{\operatorname{co} B} = e_0 \rightharpoonup H$. This implies the proposition. $\square$

1.6. One dimensional Yetter-Drinfeld modules

Let $g \in G(H)$, $\eta \in G(H^*)$. Let $V_{g,\eta}$ denote the one dimensional vector space endowed with the action $h.1 = \eta(h)1$, $h \in H$, and the coaction $1 \mapsto g \otimes 1$. The following is a restatement of a result due to Radford [**R**, Proposition 10] that describes the group-like elements in the dual of $D(H)$.

LEMMA 1.6.1. *The one-dimensional Yetter-Drinfeld modules of H are exactly of the form $V_{g,\eta}$, where $g \in G(H)$ and $\eta \in G(H^*)$ are such that $(\eta \rightharpoonup h)g = g(h \leftharpoonup \eta)$, for all $h \in H$.* $\square$

REMARK 1.6.2. Let $g \in G(H)$, $\eta \in G(H^*)$. Then $V_{g,\eta}$ is a Yetter-Drinfeld module of H if and only if $V_{\eta,g}$ is a Yetter-Drinfeld module of H^*.

PROOF. We use Lemma 1.6.1. We have that $V_{g,\eta}$ is a Yetter-Drinfeld module of H if and only if $(\eta \rightharpoonup h)g = g(h \leftharpoonup \eta)$, for all $h \in H$. This equivalent to $(g \rightharpoonup f)\eta = \eta(f \leftharpoonup g)$, for all $f \in H^*$. Hence the claim follows. $\square$

LEMMA 1.6.3. *Let $g \in G(H)$ and $\eta \in G(H^*)$ such that $V_{g,\eta}$ is a Yetter-Drinfeld module of H. Let also $q : H \to k\langle \eta \rangle$ be the Hopf algebra map obtained by transposing the inclusion $k\langle \eta \rangle \subseteq H^*$.*

Suppose that $V \subseteq H^{\operatorname{co} q}$ is a subspace of H such that $g^{-1}vg = v$, for all $v \in V$. Then $V \subseteq {}^{\operatorname{co} q}H$.

PROOF. Using Lemma 1.6.1, we have for all $v \in V$,
$$v \leftharpoonup \eta = g^{-1}(\eta \rightharpoonup v)g = g^{-1}vg = v,$$
the second equality because $v \in H^{\operatorname{co} q}$, see Remark 1.3.2. This shows that $v \in {}^{\operatorname{co} q}H$ and finishes the proof of the lemma. $\square$

The following result will be used in Chapters 6 and 12.

THEOREM 1.6.4. *Let g, η and q be as in Lemma 1.6.3. Let $A \subseteq H$ be a Hopf subalgebra such that $g^{-1}ag = a$, for all $a \in A^{\operatorname{co} q}$. Then the restriction $q|_A : A \to k\langle \eta \rangle$ is normal.*

PROOF. We shall prove that $A^{\operatorname{co} q}$ is a Hopf subalgebra of A. By Lemma 1.6.3, we have $A^{\operatorname{co} q} \subseteq {}^{\operatorname{co} q}A$. Thus ${}^{\operatorname{co} q}A = A^{\operatorname{co} q}$, since they have the same finite dimension. This implies the theorem. $\square$

1.7. $H^{co B}$ as a left coideal of H

Let $q : H \to B$ be a surjective Hopf algebra map. Identify B^* with its image under the transpose map $q^* : B^* \to H^*$; so that B^* is a Hopf subalgebra of H^*.

For each left B^*-module W, we may consider the *induced* left H^*-module $V = \operatorname{Ind}_{B^*}^{H^*} W := H^* \otimes_{B^*} W$. Most basic properties of the induction functor are discussed, for instance, in [**AN2**]. Proposition 1.7.2 below establishes a relationship between the decomposition of the induced module $\operatorname{Ind}_{B^*}^{H^*} \epsilon$ and the normality of the map q. Here, $\operatorname{Ind}_{B^*}^{H^*} \epsilon$ indicates the representation induced from the *trivial* one-dimensional representation $W = k_\epsilon$ of B^*. The key ingredient is the following identification.

Recall that $H^{co B}$ is a left coideal of H, hence a right H^*-module. Thus $(H^{co B})^*$ is naturally a left H^*-module through the action given by $\langle p.f, x \rangle := \langle p, x_1 \rangle \langle f, x_2 \rangle$, for $p \in H^*$, $f \in (H^{co B})^*$, $x \in H^{co B}$.

LEMMA 1.7.1. $(H^{co B})^* \simeq \operatorname{Ind}_{B^*}^{H^*} \epsilon$ *as left H^*-modules.*

As a consequence of this lemma, we see that for every irreducible left coideal V of H, V^* appears in $H^{co B}$ with the same multiplicity as V does.

PROOF. As left H^*-modules, $\operatorname{Ind}_{B^*}^{H^*} \epsilon = H^* \otimes_{B^*} k_\epsilon \simeq H^*/H^*(B^*)^+$. On the other hand, the evaluation map $\langle \, , \, \rangle : H^* \otimes H \to k$ induces a left H^*-linear isomorphism $H^*/H^*(B^*)^+ \simeq (H^{co B})^*$. □

PROPOSITION 1.7.2. *The map $q : H \to B$ is normal if and only if every irreducible H^*-module V appears with multiplicity $\dim V$ or 0 in $\operatorname{Ind}_{B^*}^{H^*} \epsilon$.*

PROOF. We know that q is normal if and only if $H^{co B}$ is a subcoalgebra of H. In turn, the last holds if and only if for every irreducible H-coideal $V \subseteq H^{co B}$, $H^{co B}$ contains the simple subcoalgebra corresponding to V; that is, if and only if, every irreducible left coideal of H appears with multiplicity $\dim V$ or 0 in $H^{co B}$. The proposition follows from Lemma 1.7.1. □

CHAPTER 2

The Nichols-Richmond Theorem

Recall from 1.2 that the *character algebra* of H, denoted $R(H)$, is the subalgebra of H^* spanned by the irreducible characters of H.

A subalgebra M of $R(H)$ is called a *standard* subalgebra if M is spanned by irreducible characters of H. So that if X is a subset of $\widehat{H}$, X spans a standard subalgebra of $R(H)$ if and only if the product of characters in X decompose as a sum of characters in X.

By [**NR**, Theorem 6] there is a bijection between standard subalgebras of $R(H)$ and quotient Hopf algebras of H. Under this bijection, the quotient $H \to B$ corresponds to the character algebra of B: $R(B) \subseteq R(H)$.

2.1. Irreducible characters of degree 2

In this section we collect some facts related to the fusion rules of irreducible characters of degree 2.

Suppose that H contains an irreducible character χ of degree 2, such that

$$(2.1.1) \qquad \chi^2 = \sum_{g \in G[\chi]} g;$$

in particular, $\chi^* = \chi$ and $|G[\chi]| = 4$. The set $G[\chi] \cup \{\chi\}$ spans a standard subalgebra of $R(H)$. This subalgebra corresponds to a quotient Hopf algebra $H \to \overline{H}$, where $\overline{H}$ is a semisimple non-commutative Hopf algebra of dimension 8 with algebra type $(1, 4; 2, 1)$. The classification of semisimple Hopf algebras of dimension 8 implies that the group $G[\chi] = G(\overline{H}^*)$ is not cyclic; a more general picture will appear later in 3.4.

Conversely, every semisimple non-commutative Hopf algebra of dimension 8 has four linear characters, which constitute a group isomorphic to $\mathbb{Z}_2 \times \mathbb{Z}_2$, and one (self-dual) irreducible character of degree 2, satisfying the relations (2.1.1).

Suppose now that H has an irreducible character χ of degree 2 such that

$$(2.1.2) \qquad \chi\chi^* = \epsilon + g + \chi,$$

for some $g \in G(H^*)$. Then $G[\chi] = \{\epsilon, g\}$, $\chi^* = \chi$ and $G[\chi] \cup \{\chi\}$ spans a standard subalgebra of $R(H)$, which corresponds to a quotient Hopf algebra $H \to \overline{H}$, where $\overline{H} \simeq k\mathbb{S}_3$ is the unique non-commutative semisimple Hopf algebra of dimension 6.

PROPOSITION 2.1.3. *Suppose that the following conditions are fulfilled:*

(i) $|\{\chi \in \widehat{H} : \chi(1) = 2\}|$ *is odd;*
(ii) $G(H^*)$ *contains a subgroup Γ of order 4.*

Then there is a quotient Hopf algebra $H \to \overline{H}$, where $\overline{H}$ is a semisimple non-commutative Hopf algebra of dimension 8 such that $\Gamma \subseteq \overline{H}^*$.

PROOF. The group Γ acts on the set $X_2 = \{\chi \in \widehat{H} : \chi(1) = 2\}$ in the form $g.\chi = g\chi g^{-1}$. Let $X_2' \subseteq X_2$ be the set of fixed points under this action. Since $|X_2|$ is odd by assumption, then $|X_2'|$ is also odd. Moreover, since Γ is abelian, Γ acts on X_2' by left multiplication.

Let $Y \subseteq X_2'$ be the set of fixed points of X_2' under left multiplication by Γ. Once again we find that $|Y|$ is odd and, in particular, that Y is not empty. It is easy to see that $\mu^* \in Y$ for all $\mu \in Y$, whence there must exist $\chi \in Y$ such that $\chi^* = \chi$.

By construction, $\chi^2 = \chi\chi^* = \sum_{g \in \Gamma} g$; see (1.2.3). Hence the set $\Gamma \cup \{\chi\}$ spans a standard subalgebra of $R(H)$ which corresponds to a Hopf algebra quotient of dimension 8 as claimed. □

LEMMA 2.1.4. *Suppose that $\lambda\lambda^* \in kG(H^*)$ for some irreducible character λ. Assume in addition that $\deg \lambda \leq \deg \mu$ for all irreducible character μ with $\deg \mu > 1$. Then also $\lambda^*\lambda \in kG(H^*)$.*

PROOF. By assumption we have $\lambda\lambda^* = \sum_{g \in G[\lambda]} g$. On the other hand, we may write $\lambda^*\lambda = \sum_{g \in G[\lambda^*]} g + \sum_{\deg \mu > 1} n_\mu \mu$, where $n_\mu = m(\mu, \lambda^*\lambda)$. Then we have

$$\lambda\lambda^*\lambda = \sum_{g \in G[\lambda^*]} \lambda g + \sum_{\deg \mu > 1} n_\mu \lambda\mu = |G[\lambda^*]|\lambda + \sum_{\deg \mu > 1} n_\mu \lambda\mu,$$

and also

$$\lambda\lambda^*\lambda = \sum_{g \in G[\lambda]} g\lambda = |G[\lambda]|\lambda.$$

Comparing the multiplicity of λ in both expressions, we find that $\lambda\mu = (\deg \mu)\lambda$, for all μ such that $n_\mu \neq 0$. Thus $\deg \mu = m(\lambda, \lambda\mu) = m(\mu, \lambda^*\lambda) = n_\mu$. But then $n_\mu \deg \mu \geq (\deg \lambda)^2$ for all μ such that $\deg \mu > 1$ and $n_\mu \neq 0$; so we see that $n_\mu = 0$, for all μ such that $\deg \mu > 1$. This finishes the proof of the lemma. □

2.2. The Nichols-Richmond theorem

The following theorem is due to Nichols and Richmond. See [**NR**, Theorem 11]. We state here a version convenient to the finite dimensional context.

THEOREM 2.2.1. *Suppose that H has an irreducible character χ of degree 2. Then at least one of the following conditions holds:*
(i) $G[\chi] \neq 1$;
(ii) H has a Hopf algebra quotient of dimension 24, which has a degree one character g of order 2 such that $g\chi \neq \chi$;
(iii) H has a Hopf algebra quotient of dimension 12 or 60. □

Notice that, as a consequence of the theorem, if $\dim H < 60$ and H has an irreducible character of degree 2, then $G(H^*) \neq 1$.

REMARK 2.2.2. (i) Suppose that H has an irreducible character χ of degree 2 such that $G[\chi] = 1$. Then H must also contain an irreducible character ψ of degree 3, which is necessarily self-dual, and such that $\chi\chi^* = \epsilon + \psi$.

Assume that H has no irreducible character of degree 4. Then $|G[\psi]| = 3$ and $\psi^2 = \sum_{g \in G[\psi]} g + 2\psi$; so that $G[\psi] \cup \psi$ spans a standard subalgebra of $R(H)$, which corresponds to a quotient Hopf algebra of type $(1, 3; 3, 1)$.

PROOF. We follow the lines of the proof of Case 1 in [**NR**, Theorem 11]. Since H has no irreducible character of degree 4 and $m(\chi, \psi\chi) = 1$, a counting argument implies that $\psi\chi$ is a sum of irreducible characters of degree 2. Moreover, it is easy to see that all these characters are conjugated to χ. By [**NR**, Theorem 10 (3)], $\psi\chi = \sum_{g \in G[\psi]} g\chi$, so in particular $|G[\psi]| = 3$. Multiplying both sides of this equation by χ^* on the right, we get $\psi^2 = \sum_{g \in G[\psi]} g + 2\psi$. This proves the claim. □

(ii) The assumption $G[\chi] = 1$, for an irreducible character χ of degree n, implies also that χ has exactly $|G(H^*)|$ distinct conjugates under the action of $G(H^*)$ by left multiplication. Thus, in this case, we must have an inequality $|G(H^*)| \leq |X_n|$.

The following corollary of the Nichols-Richmond Theorem gives some restrictions for the possibility $G(H^*) = 1$.

COROLLARY 2.2.3. *Suppose that H has an irreducible character of degree 2. If $G(H^*) = 1$, then H has a Hopf algebra quotient of dimension 60. In particular, $60/\dim H$.*

PROOF. Any semisimple Hopf algebra of dimension 12 contains nontrivial group-like elements [**F**]. By Theorem 2.2.1, if $G(H^*) = 1$, H must have a Hopf algebra quotient of dimension 60 as claimed. □

2.3. An application to $D(H)$

In this section, we consider an application of Theorem 2.2.1 to the Drinfeld double of H.

PROPOSITION 2.3.1. *Suppose that H is a semisimple Hopf algebra such that $\dim H < 60$. If the Drinfeld double $D(H)$ has an irreducible character of degree 2, then $G(D(H)^*) \neq 1$.*

In particular, $D(H)$ is not simple; see [**N**, Corollary 2.3.2].

PROOF. If $D(H)$ has an irreducible character of degree 2, then by Corollary 2.2.3, $G(D(H)^*) \neq 1$, unless there is a quotient Hopf algebra $q : D(H) \to B$, with $\dim B = 60$ such that $G(B^*) = 1$.

If this were the case, then by [**NZ**] $\dim H$ is divisible by 30 (since it must be divisible by 2, 3 and 5); thus by assumption, $\dim H = 30$. The proposition follows from Theorem 2, which says that H is necessarily trivial. □

In our applications we shall combine the preceding proposition with the following fact.

LEMMA 2.3.2. *Suppose that H has a Hopf subalgebra or quotient Hopf algebra of index 3. If $G(H^*) \cap Z(H^*) = 1$ and $G(H) \cap Z(H) = 1$, then the Drinfeld double of H contains an irreducible character of degree 2.*

PROOF. We may assume that H has a quotient Hopf algebra $H \to B$ of index 3; the other case is dual, once we notice that $D(H^{*\mathrm{cop}}) \simeq D(H)^{\mathrm{op}} \simeq D(H)$. By assumption $\dim H^{\mathrm{co}\,B} = 3$, and by Lemma 1.4.1 $H^{\mathrm{co}\,B}$ is a Yetter-Drinfeld submodule of H. Decomposing $H^{\mathrm{co}\,B}$ as a direct sum of irreducible Yetter-Drinfeld submodules implies the lemma, since the trivial appears with multiplicity 1. □

2.4. Existence of proper Hopf subalgebras

In this section we apply the Nichols-Richmond theorem in order to assure, in certain cases, the existence of proper Hopf subalgebras.

LEMMA 2.4.1. *Let χ and ψ be irreducible characters of H such that the product $\chi\psi$ is irreducible. Then for all irreducible character $\mu \neq \epsilon$ we must have $m(\mu, \psi\psi^*) = 0$ or $m(\mu, \chi^*\chi) = 0$.*

In particular $G[\psi] \cap G[\chi^*] = 1$.

2.4. EXISTENCE OF PROPER HOPF SUBALGEBRAS

PROOF. Let $\zeta = \chi\psi$. By Schur's Lemma, ζ is irreducible if and only if $m(\epsilon, \zeta\zeta^*) = 1$.

On the other hand, we have

$$\zeta\zeta^* = \chi\psi\psi^*\chi^* = \chi\left(\sum_{\mu \in \widehat{H}} m(\mu, \psi\psi^*)\mu\right)\chi^*$$

$$= \chi\chi^* + \sum_{\mu \neq \epsilon} m(\mu, \psi\psi^*)\chi\mu\chi^*.$$

Therefore, $m(\epsilon, \zeta\zeta^*) = 1$ if and only if for all $\mu \neq \epsilon$, with $m(\mu, \psi\psi^*) > 0$, we have $m(\epsilon, \chi\mu\chi^*) = 0$ or equivalently, $m(\mu, \chi^*\chi) = 0$. This proves the lemma. □

THEOREM 2.4.2. *Suppose that $1 \neq G[\chi] \cap G[\psi]$, for all irreducible characters χ and ψ of degree 2. Then there is a quotient Hopf algebra $\pi: H \to \overline{H}$, such that $\overline{H}$ is of type $(1, |G(H^*)|; 2, |X_2|)$ as an algebra.*

PROOF. It follows from Lemma 2.4.1 and the assumptions, that if χ and ψ are irreducible characters of degree 2, then their product $\chi\psi$ is not irreducible. Thus $\chi\psi$ decomposes as a sum of irreducible characters of degree at most 2; indeed, the one-dimensional characters appearing in $\chi\psi$ with positive multiplicity form a coset of the stabilizer $G[\chi]$ in $G(H^*)$ and thus there are 0, 2 or 4 of them, so that the other irreducible summands must be of degree 2. Therefore, the set $\{\chi \in \widehat{H} : \chi(1) \leq 2\}$ spans a standard subalgebra of the character algebra of H, implying the claim. □

We next collect some useful variations of this theorem.

REMARK 2.4.3. (i) If H has no irreducible character of degree 4 and the dimension of H is not divisible by 12, then the set $G(H^*) \cup X_2$ spans a standard subalgebra of $R(H)$. Therefore, H has a quotient Hopf algebra of type $(1, |G(H^*)|; 2, |X_2|)$.

PROOF. In this case the product of two characters of degree 2 cannot be irreducible. By Theorem 2.2.1, since 12 does not divide $\dim H$, then $G[\chi] \neq 1$ for all $\chi \in X_2$, implying that for all $\psi \in X_2$, $\chi\psi$ decomposes as a sum of irreducible characters of degree ≤ 2 (arguing as in the proof of Theorem 2.4.2). □

(ii) Suppose that $G[\chi] \neq 1$ for all $\chi \in X_2$. Assume in addition that $G(H^*)$ has a unique subgroup F of order 2. Then all irreducible characters of degree 2 are stable under multiplication by F.

In particular $F \subseteq G[\psi] \cap G[\chi^*]$, for all irreducible characters ψ and χ of degree 2.

(iii) Suppose that the action by right multiplication of $G(H^*)$ on X_2 is transitive and $|G[\chi]| \neq 1$ for some irreducible character of degree 2. Then the set $G(H) \cup X_2$ spans a standard subalgebra of $R(H^*)$ which corresponds to a quotient Hopf algebra $H \to B$, such that B is of type $(1, |G(H^*)|; 2, |X_2|)$ as an algebra.

PROOF. The assumption implies that $|G[\chi]| \neq 1$, for all irreducible character χ with $\deg \chi = 2$: indeed, the action of $G(H^*)$ by right multiplication, which is transitive, preserves $G[\chi]$ and this group is not trivial at least for one χ. Therefore $\chi\chi^*$ belongs to the span of $G(H^*)$ and X_2.

Since the action of $G(H^*)$ on the set X_2 by right multiplication is transitive, for all $\chi' \in X_2$, there exists $h \in G(H^*)$ such that $\chi' = \chi^* h$; then $\chi\chi' = \chi\chi^* h$ belongs to the span of $G(H^*) \cup X_2$.

This shows that the set $G(H^*) \cup X_2$ spans a standard subalgebra of $R(H)$ and implies the claim. □

2.5. Hopf subalgebras of index 3

Using character theory, we shall show that for some Hopf algebra inclusions $A \subseteq H$, with $[H : A] = 3$, there is a quotient Hopf algebra $H \to \overline{H}$, such that the simple $\overline{H}$-modules have dimension 1 or 2.

Observe that, by Corollary 1.4.3, if H has a Hopf subalgebra of index 3 which is not normal, then the dimension of H must be even; hence $6 / \dim H$. The following theorem and its corollary give more precise information.

THEOREM 2.5.1. *Let $A \subseteq H$ be a Hopf subalgebra such that $[H : A] = 3$. Suppose that A is not normal in H. Then there exist a Hopf algebra $\overline{H}$ of algebra type $(1, |G(H^*)|; 2, \dfrac{|G(H^*)|}{2})$ together with surjective Hopf algebra maps $H \to \overline{H} \to B$, where $B \simeq k\mathbb{S}_3$, and a surjective coalgebra map $B \to H/HA^+$ such that the following diagram is commutative:*

$$\begin{array}{ccc} H & \longrightarrow & \overline{H} \\ \downarrow & & \downarrow \\ H/HA^+ & \longleftarrow & B. \end{array}$$

PROOF. Since A is not normal in H, by Proposition 1.7.2, $\operatorname{Ind}_A^H \epsilon = \epsilon + \chi$, where χ is an irreducible character of degree 2. In particular, $\chi^* = \chi$, by Lemma 1.7.1.

CLAIM 2.5.2. *$|G[\chi]| = 2$ and $G[\chi] \cup \{\chi\}$ spans a standard subalgebra of $R(H)$.*

2.5. HOPF SUBALGEBRAS OF INDEX 3

PROOF. It is enough to show that the product $\chi\chi^* = \chi^2$ admits a decomposition as in (2.1.2).

Since $m(\chi, \mathrm{Ind}_A^H \epsilon) = 1$, we have $\chi|_A = \epsilon + x$, where $\epsilon \neq x \in G(A^*)$, by Frobenius reciprocity. Hence $\chi^2|_A = 2(\epsilon + x)$.

In the character algebra of H one of the following decompositions must hold:

(a) $\chi^2 = \epsilon + a + b + c$, where $a, b, c \in G(H^*) \backslash \{\epsilon\}$ are pairwise distinct;
(b) $\chi^2 = \epsilon + \psi$, where ψ is an irreducible character of degree 3;
(c) $\chi^2 = \epsilon + a + \lambda$, where $\epsilon \neq a \in G(H^*)$ and λ is an irreducible character of degree 2.

We shall show that the cases (a) and (b) are impossible, thus proving that $|G[\chi]| = 2$. Suppose that (b) holds. Restricting to A, we find that necessarily $m(\epsilon, \psi|_A) > 0$. But by Frobenius reciprocity $m(\epsilon, \psi|_A) = m(\psi, \mathrm{Ind}_A^H \epsilon) = 0$; this contradiction discards case (b). Case (a) is similarly discarded.

Hence (c) holds. It remains to show that $\lambda = \chi$. For this, we restrict the equation $\chi^2 = \epsilon + a + \lambda$ to A, and apply the Frobenius reciprocity to find that $m(\lambda, \mathrm{Ind}_A^H \epsilon) > 0$, whence $\lambda = \chi$. This proves the claim. □

Clearly, the standard subalgebra in Claim 2.5.2 corresponds to a quotient Hopf algebra $H \to B$, such that $B \simeq k\mathbb{S}_3$. Moreover, by construction, and using Lemma 1.7.1, $(H^*)^{\mathrm{co}\,A^*} \subseteq B^* \subseteq H^*$. Hence there is a surjective coalgebra map $B \to H/HA^+$ which factorizes the canonical map $H \to H/HA^+$.

Lemma 1.7.1 implies also that $g\chi g^{-1} = \chi$, for all $g \in G(H^*)$. Therefore, $G(H^*) \cup G(H^*)\chi$ spans a standard subalgebra of $R(H)$, which corresponds to a quotient Hopf algebra $H \to \overline{H}$ with the desired properties. This finishes the proof of the theorem. □

Note that $\dim \overline{H} = 3|G(H^*)|$. The following corollary is an immediate consequence of the theorem.

COROLLARY 2.5.3. *Suppose that $A \subseteq H$ is a Hopf subalgebra such that $[H : A] = 3$ and A is not normal in H. Then we have*
(i) $|G(H^*)|$ *is even;*
(ii) $3|G(H^*)|$ *divides* $\dim H$. □

CHAPTER 3

Quotient Coalgebras

Let H be a semisimple Hopf algebra and let $A \subseteq H$ be a Hopf subalgebra. Consider the quotient coalgebra $p : H \to \overline{H} := H/HA^+$. By [**M2**, 3.4], $\overline{H}$ is a cosemisimple coalgebra. In this chapter we aim to relate the corepresentations of H and $\overline{H}$. We discuss the corepresentation theory of $\overline{H}$ in relation with that of H and the corestriction functor ${}^H\mathcal{M} \to {}^{\overline{H}}\mathcal{M}$.

We show that if C is a simple subcoalgebra of H such that $Ca \subseteq C$, for all $a \in A$, then the dual of the quotient coalgebra C/CA^+ and the crossed product A_α, where $\alpha : A \otimes A \to k$ is a certain 2-cocycle, constitute a commuting pair in C^*. This is applied in combination with Masuoka's main result in [**M5**], in some instances of the proof of Theorem 1.

In particular, when $A = kG$ is the group algebra of a subgroup G of $G(H)$ and V is a simple H-comodule, we deduce that $\operatorname{End}^{\overline{H}}(V)$ is isomorphic as an algebra to a twisted group algebra $k_\alpha \Gamma$, where $\Gamma \subseteq G$ is the stabilizer of V, i.e. $\Gamma = \{g \in G : V \otimes g \simeq V\}$, and $\alpha : \Gamma \times \Gamma \to k^\times$ is a 2-cocycle. This result implies that the multiplicity of an irreducible $\overline{H}$-comodule in V is a divisor of the order of Γ. In particular, when the group Γ is abelian, all irreducible $\overline{H}$-comodules in the restriction of V to $\overline{H}$ appear with the same multiplicity d, where d divides the order of Γ. This allows us to recover the result in [**M5**, Proposition 2.4].

Some of these results are applied to the case when H is a biproduct in the sense of Radford: $H \simeq R\#A$. Indeed, in this case R is isomorphic as a coalgebra to the quotient H/HA^+.

We shall denote by ${}^H\mathcal{M}$ the category of left H-comodules; for a left H-comodule algebra A, the category of (left-right) (A, H)-Hopf modules will be indicated by ${}^H\mathcal{M}_A$.

3.1. A multiplicity formula

Let $\rho : V \to H \otimes V$, $\rho(v) = v_{-1} \otimes v_0$, be a left H-comodule. Consider the $\overline{H}$-comodule structure on V obtained by corestriction along the coalgebra map p; we shall sometimes use the notation $\overline{V}$ for this comodule structure, as well as $\overline{\rho} = (p \otimes \mathrm{id})\rho : V \to \overline{H} \otimes V$ for the structure map. We obtain in this way a functor ${}^H\mathcal{M} \to {}^{\overline{H}}\mathcal{M}$, $V \mapsto \overline{V}$.

By a result of Schneider [**Sc3**, Theorem II], there is a category equivalence $\omega : {}^H\mathcal{M}_A \to {}^{\overline{H}}\mathcal{M}$ between the category of (A, H)-Hopf modules and the category of $\overline{H}$-comodules. The equivalence is given by $\omega : M \mapsto M/MA^+$, for any object M of ${}^H\mathcal{M}_A$.

Consider now the functor $F : {}^H\mathcal{M} \to {}^H\mathcal{M}_A$, $F(V) = V \otimes A$; where H coacts diagonally on $V \otimes A$ and the right A-module structure is given by $v \otimes a.b = v \otimes ab$.

LEMMA 3.1.1. *(i) The functor F is a left adjoint of the forgetful functor $U : {}^H\mathcal{M}_A \to {}^H\mathcal{M}$;*
(ii) for all left H-comodules V there is an isomorphism $\omega F(V) \simeq \overline{V}$.

PROOF. (i) We define natural maps
$$\psi : \mathrm{Hom}^H(X, U(Y)) \to \mathrm{Hom}^H_A(X \otimes A, Y),$$
$$\phi : \mathrm{Hom}^H_A(X \otimes A, Y) \to \mathrm{Hom}^H(X, U(Y)),$$
as follows:
$$\psi(f)(x \otimes a) = f(x).a, \quad \phi(g)(x) = g(x \otimes 1),$$
for all $x \in X$, $a \in A$. It is not hard to check that ψ and ϕ are well defined and are indeed inverse isomorphisms.

(ii) The maps $\mu : (V \otimes A)/(V \otimes A^+) \to \overline{V}$, $\mu([v \otimes a]) := \epsilon(a)v$, and $\eta : \overline{V} \to (V \otimes A)/(V \otimes A^+)$, $\eta(v) := [v \otimes 1]$, define inverse $\overline{H}$-colinear isomorphisms. □

As a consequence we get the following proposition.

PROPOSITION 3.1.2. *Suppose that U and V are finite-dimensional left H-comodules. There is a natural linear isomorphism*
$$\mathrm{Hom}^{\overline{H}}(\overline{U}, \overline{V}) \simeq \mathrm{Hom}^H(V^* \otimes U, A).$$
Suppose that V is a simple H-comodule. Then $\overline{V}$ is simple if and only if
$$\mathrm{Hom}^H(V, V \otimes W) = \mathrm{Hom}^H(V^* \otimes V, W) = 0,$$
for all simple left A-comodule $k1 \neq W$.

PROOF. We have isomorphisms
$$\mathrm{Hom}^H(U, V \otimes A) \simeq \mathrm{Hom}^H_A(U \otimes A, V \otimes A)$$
$$\simeq \mathrm{Hom}^{\overline{H}}(\omega(U \otimes A), \omega(V \otimes A)) \simeq \mathrm{Hom}^{\overline{H}}(\overline{U}, \overline{V});$$

the first isomorphism by Lemma 3.1.1 (i), the second since ω is a category equivalence and the third by Lemma 3.1.1 (ii). This proves the first statement since $\operatorname{Hom}^H(U, V \otimes A) \simeq \operatorname{Hom}^H(V^* \otimes U, A)$.

As left H-comodule, A decomposes in the form $A \simeq \oplus_W (\dim W) W$, where the sum runs over the set of isomorphism classes of irreducible left coideals of A, which coincides with the set of isomorphism classes of irreducible left coideals W of H which are contained in A. Hence we have

$$\operatorname{Hom}^{\overline{H}}(\overline{V}, \overline{V}) \simeq \operatorname{Hom}^H(V^* \otimes V, A) \simeq \oplus_W (\dim W) \operatorname{Hom}^H(V^* \otimes V, W);$$

this implies the last statement, in view of Schur's Lemma. □

REMARK 3.1.3. (i) Suppose that U and V are finite-dimensional left H-comodules, and let χ_U and $\chi_V \in H$ be the corresponding characters. As a left H-comodule, $A \simeq \oplus_{\lambda \in \widehat{A^*}} \deg \lambda \, W_\lambda$. Proposition 3.1.2 implies the following multiplicity formula:

$$\dim \operatorname{Hom}^{\overline{H}}(\overline{U}, \overline{V}) = \sum_{\lambda \in \widehat{A^*}} \deg \lambda \, m(\lambda, \chi_V^* \chi_U).$$

(ii) Suppose that $A = kG$, where G is a subgroup of $G(H)$. Let $G[V^*]$ denote the subgroup of G consisting of all elements g for which $Vg \simeq V$; that is, $G[V^*] = G \cap G[\chi_V^*]$. Recall from (1.2.2) that for $g \in G$, we have $\dim \operatorname{Hom}^H(V^* \otimes V, g) = 1$ if and only if $g \in G[V^*]$, and $\dim \operatorname{Hom}^H(V^* \otimes V, g) = 0$ otherwise.

It follows from Proposition 3.1.2 that

$$\operatorname{End}^{\overline{H}} V \simeq \operatorname{Hom}^H(V^* \otimes V, kG) = \bigoplus_{g \in G[V^*]} \operatorname{Hom}^H(V^* \otimes V, g).$$

We thus get $\dim \operatorname{End}^{\overline{H}} V = |G[V^*]|$.

3.2. Stable subcoalgebras

We keep the notation in the previous section. Let $C \subseteq H$ be a simple subcoalgebra and let $V \subseteq C$ be an irreducible left coideal of H. Suppose that $Ca \subseteq C$, for all $a \in A$. That is, C is a right A-module with action given by right multiplication; by [**NZ**] C is a free right A-module. Let $\overline{C} := p(C)$ and let $t \in A$ be the normalized integral. We have $H = H(1-t) \oplus Ht$, and $HA^+ = H(1-t)$. In particular, $p(h) = p(ht)$, for all $h \in H$.

Notice also that $\operatorname{End}^H V = \operatorname{End}^C V$ and $\operatorname{End}^{\overline{H}} V = \operatorname{End}^{\overline{C}} V$.

LEMMA 3.2.1. *(i) The map $p : H \to \overline{H}$ induces an identification $C/C(1-t) = \overline{C}$;*
(ii) $\dim \overline{C} = (\dim A)^{-1} \dim C$.

PROOF. (i) The map p induces an identification $\overline{C} = C/C \cap HA^+ = C/C \cap H(1-t)$. We claim that $C \cap H(1-t) = C(1-t)$. Indeed, let $c \in C \cap H(1-t)$; then $c = h(1-t)$ implying, since $(1-t)$ is an idempotent, that $c = c(1-t) \in C(1-t)$. The other inclusion is immediate from the fact that $Ca \subseteq C$, for all $a \in A$.

(ii) We have $C = C(1-t) \oplus Ct$, since t and $1-t$ are orthogonal idempotents, and C is stable under right multiplication by A. By part (i), $\dim \overline{C} = \dim Ct$. But $Ct = C^A$ is the space of A-invariant elements in C under the action by right multiplication. Since C is a free right A-module of rank $(\dim A)^{-1} \dim C$, then $\dim C^A = (\dim A)^{-1} \dim C$. □

REMARK 3.2.2. Let $\chi \in C$ be the irreducible character of H corresponding to C. The simple subcoalgebra C satisfies $Ca \subseteq C$ for all $a \in A$ if and only if $\chi \psi \in \mathbb{Z}\chi$, for all $\psi \in \widehat{A^*}$. This follows from the fact that the multiplication map $m : H \otimes H \to H$ is a left (and right) H-comodule map.

Since C is a right A-module coalgebra under right multiplication, then C^* is a left A-module algebra under the action $(a.f)(c) = f(ca)$, $f \in C^*$, $c \in C$, $a \in A$.

By the Skolem-Noether Theorem for Hopf algebras [**M9**], since C^* is a simple algebra over k, there exists a convolution invertible map $\psi : A \to C^*$ such that $a.f = \psi(a_1) f \psi^{-1}(a_2)$, for all $f \in C^*$, $a \in A$. This gives rise to an algebra map $\psi : A_\alpha \to C^*$, where $\alpha \in Z^2(A, k)$ is the 2-cocycle associated to ψ in the form $\alpha(a, b) = \psi^{-1}(a_1 b_1) \psi(a_2) \psi(b_2)$. Here, and elsewhere, $A_\alpha := A \#_\alpha k$ denotes the associated crossed product with respect to the cocycle α; that is, $A_\alpha = A$ as vector spaces, with the multiplication $a.b = \alpha(a_2, b_2) a_1 b_1$, $a, b \in A$. Note that, when $A = kG$ is a group algebra, $(kG)_\alpha = k_\alpha G$ is the *twisted* group algebra.

PROPOSITION 3.2.3. *As algebras,* $\left(\mathrm{End}^{\overline{C}} V\right)^{\mathrm{op}} \simeq A_\alpha$. *Moreover, A_α and $\overline{C}^*$ form a commuting pair in* $(\mathrm{End}\, V)^{\mathrm{op}} \simeq C^*$.

PROOF. The coalgebra projection $p : C \to \overline{C}$ induces by transposition an algebra inclusion $\overline{C}^* \subset C^*$. We have that $\overline{C}^*$ coincides with the subalgebra of invariants $(C^*)^A$ under the action of A. Indeed, $f \in \overline{C}^*$ if and only if $f(c) = f(p(c))$, for all $c \in C$, if and only if $f(c) = f(ct) = (t.f)(c)$, for all $c \in C$; whence, $\overline{C}^* = (C^*)^A$.

Hence, by definition of ψ, $\overline{C}^*$ coincides with the commutant of $\psi(A_\alpha)$ in C^*. Since $\overline{C}^*$ is semisimple, the double commutant theorem implies that $\psi(A_\alpha)$ is the commutant of $\overline{C}^*$ in C^*.

3.2. STABLE SUBCOALGEBRAS

It follows from Burnside's Density Theorem [**CR**] that there is an anti-isomorphism of algebras $I : C^* \to \operatorname{End} V$, given by

$$(3.2.4) \qquad I(f)(v) := \langle f, v_{-1}\rangle v_0, \quad v \in V, f \in C^*.$$

Under this identification, the commutant of $\overline{C}^*$ in C^* coincides with the subalgebra $\left(\operatorname{End}^{\overline{C}} V\right)^{\operatorname{op}}$ of $(\operatorname{End} V)^{\operatorname{op}}$. Therefore ψ defines a surjective algebra map $\psi : A_\alpha \to \psi(A_\alpha) = \left(\operatorname{End}^{\overline{C}} V\right)^{\operatorname{op}}$.

Since $CA \subseteq C$, we have $\chi_V \lambda = (\deg \lambda)\chi_V$, for all irreducible character $\lambda \in \widehat{A^*}$; that is, $m(\lambda, \chi_V^* \chi_V) = \deg \lambda$, for all $\lambda \in \widehat{A^*}$. By Remark 3.1.3 (i), $\dim \operatorname{End}^{\overline{C}} V = \sum_{\lambda \in \widehat{A^*}} (\deg \lambda)^2 = \dim A$. Therefore $\psi : A_\alpha \to C^*$ is an injective algebra map and determines an isomorphism $A_\alpha \simeq \left(\operatorname{End}^{\overline{C}} V\right)^{\operatorname{op}}$. This finishes the proof of the proposition. □

COROLLARY 3.2.5. *There exists a bijective correspondence between irreducible A_α-modules and irreducible $\overline{C}$-comodules.*

There is an isomorphism $V \simeq \bigoplus_i U_i \otimes W_i$, where W_i runs over a system of representatives of the isomorphism classes of irreducible A_α-modules and U_i is the irreducible $\overline{C}$-comodule corresponding to W_i. □

In particular, if $A = kG$ where G is a finite group, then the multiplicity of U_i in V divides the order of G, for all i.

If G is an abelian group, then all the irreducible $k_\alpha G$-modules have the same dimension d. Therefore all irreducible $\overline{C}$-subcomodules of V appear with the same multiplicity d.

As an application of the methods of this section, we have the following proposition.

PROPOSITION 3.2.6. *Suppose that $AC = C = CA$. Assume in addition that $\dim A = \dim C$. Then A is normal in $k[C]$.*

Here, $k[C]$ denotes the subalgebra generated by C; this is a Hopf subalgebra of H containing A.

PROOF. By Lemma 3.2.1, $\dim Ct = \dim tC = 1$, where $t \in A$ is the normalized integral. Therefore $Ct = tC = k\psi$, where $\psi \in C$ is the corresponding irreducible character. Hence, for all $c \in C$, $tc = l(c)\psi$ and $ct = r(c)\psi$, which implies that $tc = ct$, after applying the counit.

Hence t commutes pointwise with C; thus it is central in $k[C]$, and a fortiori, A is normal in $k[C]$. □

REMARK 3.2.7. It follows from Corollary 3.2.5 that, if G is cyclic and $|G| = \dim V$, then $\overline{C}$ is a cocommutative coalgebra. We thus recover a fact in the proof of [**M5**, Proposition 2.4].

3.3. Quotients modulo group-like Hopf subalgebras

Let G be a subgroup of $G(H)$ and let $A = kG$. We shall now specialize the description in the previous sections.

Let $t := \dfrac{1}{|G|} \sum_{g \in G} g$ be the normalized integral in kG.

LEMMA 3.3.1. *Let C and D be simple subcoalgebras of H. Then the following are equivalent:*
(i) $p(C) \cap p(D) \neq 0$;
(ii) $p(C) = p(D)$;
(iii) There exists $g \in G$ such that $Cg = D$.

PROOF. (ii) $\Longrightarrow$ (i). Clear, since $p(C) \neq 0$ for all simple subcoalgebra C.

(iii) $\Longrightarrow$ (ii). If $D = Cg$, then $Dt = Ct$ and therefore $p(D) = p(Dt) = p(Ct) = p(C)$.

(i) $\Longrightarrow$ (iii). Suppose that $c \in C$ and $d \in D$ are such that $p(c) = p(d) \neq 0$. Then $c - d$ belongs to the kernel of p, and there exists $h \in H$ such that $c - d = h(1 - t)$. Hence
$$(c - d)t = h(1 - t)t = 0,$$
implying that $ct = dt$. But $ct \in \sum_{g \in G} Cg$ and $dt \in \sum_{g \in G} Dg$. Therefore, $Cg \cap Dg' \neq 0$, for some $g, g' \in G$; since both Cg and Dg' are simple subcoalgebras, this implies that $Cg = Dg'$ and $D = Cg(g')^{-1}$. □

The group G acts on the set of simple subcoalgebras of H by right multiplication. Let $C_1, \ldots, C_n$ be a system of representatives of this action, and let $G_i \subseteq G$ be the stabilizer of C_i.

COROLLARY 3.3.2. *There is an isomorphism of coalgebras $\overline{H} \simeq \bigoplus_{i=1}^n \overline{C_i}$, where $\overline{C_i} = p(C_i) \simeq C_i/C_i(kG_i)^+$.*

PROOF. It follows from Lemma 3.3.1 that $\overline{H} \simeq \bigoplus_{i=1}^n p(C_i)$. Thus it remains to see that $p(C_i) \simeq C_i/C_i(kG_i)^+$.

Fix $1 \leq i \leq n$ and let $C = C_i$, $G_C = G_i$. We claim that $C \cap H(kG)^+ = C \cap C(1 - t) = C \cap C(1 - t_C)$, where $t_C = \dfrac{1}{|G_C|} \sum_{h \in G_C} h$ is the normalized integral in kG_C.

Note first that if $c \in C \cap H(kG)^+ = C \cap H(1-t)$ then $c = h(1-t)$ and thus $c(1-t) = h(1-t)^2 = c$; this shows that $C \cap H(kG)^+ = C \cap C(1-t)$. On the other hand, we have

$$t = \frac{1}{|G|} \sum_{g \in G} g = \frac{1}{|G|} \sum_{g \in G_C \backslash G} \sum_{h \in G_C} hg = \frac{|G_C|}{|G|} \sum_{g \in G_C \backslash G} t_C g.$$

Thus, for $c \in C$, $ct = \frac{|G_C|}{|G|} \sum_{g \in G_C \backslash G} ct_C g$ belongs to $\bigoplus_{g \in G_C \backslash G} Cg$, implying that $ct = 0$ if and only if $ct_C = 0$. Then $c \in C \cap C(1-t)$ if and only if $ct = 0$ if and only if $ct_C = 0$ if and only if $c \in C \cap C(1-t_C) = C(1-t_C)$. This proves the claim and the corollary follows. $\square$

3.4. On the structure of $G(H)$

In this section we present some consequences of Proposition 3.2.3. Some of them are special cases of results in the papers [**TY, T**]. We keep the notation in the previous sections: $C \subseteq H$ is a simple subcoalgebra, $V \subseteq C$ is an irreducible left coideal, G is a subgroup of $G(H)$ such that $Cg = C$, for all $g \in G$, and $\overline{C} = C/C(kG)^+$ is the quotient coalgebra.

PROPOSITION 3.4.1. *Suppose that $\overline{C}$ is a simple coalgebra. Then there exists a non-degenerate 2-cocycle $\alpha : G \times G \to k^\times$. In particular, the group G is solvable and not cyclic.*

PROOF. By Corollary 3.2.5 the twisted group algebra $k_\alpha G$ is simple. This implies the proposition. $\square$

COROLLARY 3.4.2. *Suppose that $|G| = \dim C = (\dim V)^2$. Then G is solvable and not cyclic.* $\square$

REMARK 3.4.3. Suppose as above that $|G| = \dim C$. If we assume in addition that $V^* \simeq V$, then $A = kG \oplus C$ is a Hopf subalgebra whose category $\mathcal{C}$ of corepresentations has necessarily the fusion rules in [**TY**]. The results in *loc. cit.* imply that G is abelian. Since by definition $\mathcal{C}$ admits a fiber functor, the existence of a non-degenerate 2-cocycle on G is a consequence of [**T**].

In the following proposition we give a Hopf theoretical proof of these facts concerning G, under rather less restrictive assumption.

PROPOSITION 3.4.4. *Suppose that the following conditions hold:*
(i) $|G| = \dim C$;
(ii) $gC = C = Cg$, for all $g \in G$.
Then the group G is abelian and admits a non-degenerate 2-cocycle.
If in addition $C = \mathcal{S}(C)$, then $A = kG \oplus C$ is a Hopf subalgebra of H of dimension $2 \dim C$, which fits into a cocentral extension $1 \to k^{\widehat{G}} \to A \to k\mathbb{Z}_2 \to 0$.

An analogous result, in the context of Kac algebras, appears in [**IK**, Theorem IX.8 (ii)].

PROOF. Keep the notation in the proof of Proposition 3.2.3. Let $X_g \in C^*$ denote the image of $g \in G$ under the map $\psi : G \to C^*$; so that $\{X_g : g \in G\}$ is a basis of C^* and $X_a X_b = \alpha(a,b) X_{ab}$, for all $a, b \in G$. Also, the action $(a.p)(c) = p(ca)$ is given by $a.p = X_a p X_a^{-1}$, for all $p \in C^*$, $a \in G$.

In view of condition (ii), the same arguments apply to the Hopf algebra H^{op}. Therefore there exists a basis $\{Y_g : g \in G\}$ of C^* such that the action $(p.b)(c) = p(bc)$ is given by $p.b = Y_b^{-1} p Y_b$, for all $p \in C^*$, $b \in G$. The relation $(a.p).b = a.(p.b)$ implies that $X_a Y_b X_a^{-1} Y_b^{-1} = \zeta(a,b) 1$, for some map $\zeta : G \times G \to k^\times$. The definition of ζ implies that $a.Y_b = \zeta(a,b) Y_b$ and similarly that $X_a^{-1}.b = \zeta(a,b) X_a^{-1}$. Thus, by the associativity of the actions, we get that ζ is a bicharacter on G.

Suppose that there exists $a \in G$ such that $\zeta(a,b) = 1$, for all $b \in G$. This implies that $a.Y_b = Y_b$, for all $b \in G$. Since $\{Y_b\}_{b \in G}$ is a basis of C^*, we get that the action of a on C^* (and thus on C) is trivial. This implies that $a = 1$, because by [**NZ**] $C^* \simeq kG$ as a left and right kG-module. Therefore the bicharacter $\zeta : G \times G \to k^\times$ is non-degenerate. This proves that G is abelian as claimed. Finally, since kG, which is isomorphic to $k^{\widehat{G}}$, has index 2 in A, the last part of the proposition follows; see Corollary 1.4.3. □

3.5. A criterion of normality

We review in this section a result of A. Masuoka, which appears in [**M5**, Section 2].

We shall assume that $C \subseteq H$ is a simple subcoalgebra of dimension n^2 of H, and $g \in G(H)$ is a group-like element of order n such that $gC = C = Cg$. In particular, $g \in k[C]$, hence $k\langle g \rangle \subseteq k[C]$.

Let $V \subseteq C$ be an irreducible left coideal, so that $kg \otimes V \simeq V \simeq V \otimes kg$. Let also $\alpha : kg \otimes V \to V$ and $\alpha' : V \otimes kg \to V$ be H-colinear isomorphisms.

LEMMA 3.5.1. *$k\langle g \rangle$ is a normal Hopf subalgebra of $k[C]$ if and only if α and α' commute as endomorphisms of V.*

PROOF. Let $t \in k\langle g \rangle$ be the normalized integral. Then $k\langle g \rangle$ is normal in $k[C]$ if and only if $tc = ct$, for all $c \in C$. By [**M5**, Lemma 2.1] this is in turn equivalent to $[\alpha, \alpha'] = 0$ in $\mathrm{End}\, V$. □

COROLLARY 3.5.2. *Suppose that V is an irreducible left coideal of C such that $gV = V = Vg$. Then $k\langle g \rangle$ is normal in $k[C]$.*

PROOF. In this situation, we may take as α the left multiplication by g, and α' the right multiplication by g. Since this endomorphisms commute with each other, the corollary follows from Lemma 3.5.1. □

CHAPTER 4

Braided Hopf Algebras

Let A be a semisimple Hopf algebra and let ${}^A_A\mathcal{YD}$ denote the braided category of Yetter-Drinfeld modules over A. Let R be a semisimple braided Hopf algebra in ${}^A_A\mathcal{YD}$. The results in this chapter concern the *biproduct* construction, as described in Section 4.1. This construction was introduced by Radford [**R2**] and interpreted in categorical terms by Majid [**Mj, Mj2**].

4.1. Radford-Majid biproduct construction

Denote by $\rho : R \to A \otimes R$, $\rho(a) = a_{-1} \otimes a_0$, and $.: A \otimes R \to R$, the coaction and action of A on R, respectively. So that the Yetter-Drinfeld compatibility condition reads as follows:

(4.1.1) $\qquad \rho(h.a) = h_1 a_{-1} \mathcal{S}(h_3) \otimes h_2.a_0, \quad \forall a \in R,\ h \in A.$

We shall use the notation $\Delta_R(a) = a^1 \otimes a^2$ and $\mathcal{S}_R$ for the comultiplication and the antipode of R, respectively.

Thus our assumption amounts to the following conditions:

(4.1.2) R is an A-module and A-comodule algebra;

(4.1.3) R is an A-module and A-comodule coalgebra;

(4.1.4) $\Delta_R(ab) = a^1((a^2)_{-1}.b^1) \otimes (a^2)_0 b^2$;

(4.1.5) $\mathcal{S}_R(a^1)a^2 = \epsilon_R(a)1_R = a^1 \mathcal{S}_R(a^2)$.

Let $H = R\#A$ be the corresponding biproduct; so that H is a semisimple Hopf algebra with multiplication, comultiplication and antipode given by
(4.1.6)
$$(a\#g)(b\#h) = a(g_1.b)\#g_2 h, \quad \Delta(a\#g) = a^1\#(a^2)_{-1}g_1 \otimes (a^2)_0\#g_2,$$

$$\mathcal{S}(a\#g) = (1\#\mathcal{S}(a_{-1}g))(\mathcal{S}_R(a_0)\#1),$$

for all $g, h \in A$, $a, b \in R$; here we use the notation $a\#g$ to indicate the element $a \otimes g \in R\#A$. See [**R2**].

Consider the natural maps $\pi : H \to A$, $\pi(r\#a) = \epsilon_R(r)a$, and $\iota : A \to H$, $\iota(a) = 1 \otimes a$. Then π is a Hopf algebra surjection and ι is a Hopf algebra injection. Moreover we have $\pi\iota = \mathrm{id}_A$ and

(4.1.7) $$R = H^{\mathrm{co}\,\pi} = \{h \in H : (\mathrm{id}\otimes\pi)\Delta(h) = h \otimes 1\},$$

coincides, as an A-module and A-comodule algebra with the left coideal subalgebra of right A-coinvariants in H. On the other hand, the map $\mathrm{id}\otimes\epsilon : H \to R$ induces an isomorphism of left A-module and A-comodule coalgebras

(4.1.8) $$R \simeq H/HA^+.$$

Indeed, the biproduct construction for finite dimensional Hopf algebras is characterized by these properties. Namely, suppose that there are Hopf algebra maps $\iota : A \to H$ and $\pi : H \to A$ such that $\pi\iota : A \to A$ is an isomorphism. Then the subalgebra $R := H^{\mathrm{co}\,\pi}$ of right coinvariants of π has a natural structure of Yetter-Drinfeld Hopf algebra over A such that the multiplication map $R\#A \to H$ induces a Hopf algebra isomorphism. This principle will be often used throughout this paper.

Typically, and mainly following the lines described in Chapter 2, we shall encounter a Hopf subalgebra $A \subseteq H$ and a surjective Hopf algebra map $\pi : H \to B$, where H, A and B are certain semisimple Hopf algebras such that $\dim A = \dim B$. After an analysis of the possible left coideal decompositions of $H^{\mathrm{co}\,B}$, we shall be able sometimes to deduce that $A \cap H^{\mathrm{co}\,B} = k1$: this property guarantees the injectivity of $\pi|_A$, and hence that $\pi|_A : A \to B$ is an isomorphism. This will tell us that $H \simeq R\#A$ has the structure of a biproduct and will enable us to use the biproduct techniques that we discuss in the rest of this chapter.

One simple instance of this situation, frequently used along this paper, is described in the following lemma.

LEMMA 4.1.9. *Suppose $A \subseteq H$ is a cocommutative Hopf subalgebra and $\pi : H \to B$ is a surjective Hopf algebra map, such that $\dim A = \dim B$ and $\dim A$ is relatively prime to $[H : A]$. Then $\pi|_A : A \to B$ is an isomorphism, and $H \simeq R\#A$ is a biproduct.*

PROOF. It is enough to show that $A \cap H^{\mathrm{co}\,\pi} = k1$. This follows from the assumptions, since $\dim A \cap H^{\mathrm{co}\,\pi}$ divides both $\dim A$ and $\dim H^{\mathrm{co}\,\pi} = [H : A]$. $\square$

4.2. Coalgebra structure of R

It turns out that R is a normal left coideal subalgebra of H as well as a quotient left H-module coalgebra through the identification $R \simeq H/HA^+$. Since H is also cosemisimple, R decomposes as a direct sum $R = \oplus_i V_i$, where V_i are irreducible left coideals of H. It is clear that any left coideal V of H such that $V \subseteq R$ is an A-subcomodule of R.

The following lemma gives insight into the relationship between the H-comodule structure and the coalgebra structure on R.

PROPOSITION 4.2.1. *Let $V \subseteq R$ be a left coideal of H. Then the following hold:*

(i) V is a left coideal of R;

(ii) $\dim \operatorname{End}^R(V) \leq \dim V$, and the equality holds if and only if V is a subcoalgebra of R. If this is the case, and if V is an irreducible left coideal of H, then V has multiplicity 1 as a left H-subcomodule of R.

PROOF. (i) The map $p = \operatorname{id} \otimes \epsilon : H \to R$ is a coalgebra surjection, and $p|_R = \operatorname{id}_R$. Since $\Delta(V) \subseteq H \otimes V$, then $\Delta_R(V) = (p \otimes p)\Delta(V) \subseteq R \otimes V$, showing that V is a left coideal of R.

(ii) Let $V = \oplus_i m_i V_i$, where V_i are simple, pairwise non-isomorphic, left coideals of R. Thus $m_i \leq \dim V_i$, and $\dim \operatorname{End}^R(V) = \sum_i m_i^2 \leq \dim V$. Moreover, $\dim \operatorname{End}^R(V) = \dim V$ if and only if $m_i = \dim V_i$, for all i, if and only if V is a subcoalgebra of R.

Suppose now that V and U are irreducible coideals of H such that V is a subcoalgebra of R and $V \simeq U \subseteq R$. Then $U \simeq V$ as left coideals of R. Since V is a subcoalgebra of R, this implies that $U = V$. The proof of the proposition is now complete. $\square$

COROLLARY 4.2.2. *Suppose that $V \subseteq R$ is a left coideal of H and let χ be the character of V. Then we have*

$$\sum_{\lambda \in \widehat{A^*}} \deg \lambda \ m(\lambda, \chi_V^* \chi_V) \leq \dim V,$$

and the equality holds if and only if V is a subcoalgebra of R.

In particular, $|G(A) \cap G[\chi^*]| \leq \dim V$.

Recall that $G[\chi^*] = \{g \in G(H) : \chi g = \chi\}$.

PROOF. Combine Proposition 4.2.1 (ii) with Remark 3.1.3. $\square$

REMARK 4.2.3. Suppose that $A = kG$ is a group algebra and $V \subseteq R$ is an irreducible left coideal of H such that $\dim V = |G|$ and $Vg \simeq V$, for all $g \in G$, as in [**N2**, 1.3].

Then we have $\dim \operatorname{End}^R V = |G|$ and, by Proposition 3.2.3, there is an isomorphism of algebras $(\operatorname{End}^R V)^{\operatorname{op}} \simeq k_\alpha G$ for some $\alpha \in Z^2(G, k^\times)$.

On the other hand, by Proposition 4.2.1, V is a subcoalgebra of R. So that $\operatorname{End}^R V = \operatorname{End}^V V \simeq (V^*)^{\operatorname{op}}$ as algebras. Whence V is isomorphic to a dual twisted group algebra as a coalgebra. We thus recover the statement in [**N2**, Proposition 1.3.1].

4.3. Hopf subalgebras

In this section we discuss some results on the existence of proper Hopf subalgebras in the biproduct $H \simeq R\#A$.

LEMMA 4.3.1. *Suppose that $\widetilde{R} \subseteq R$ is a subspace such that:*
(1) $\widetilde{R}$ is a subalgebra and a subcoalgebra of R;
(2) there exists a Hopf subalgebra $B \subseteq A$ such that $\rho(\widetilde{R}) \subseteq B \otimes \widetilde{R}$ and $B.\widetilde{R} \subseteq \widetilde{R}$.
Then $\widetilde{R}$ is a braided Hopf algebra over B and the biproduct $\widetilde{R}\#B$ is a Hopf subalgebra of H.

PROOF. It is not hard to see that the conditions (4.1.2), (4.1.3) and (4.1.4) are verified, so that $\widetilde{R}$ is a braided Hopf subalgebra of R. Using (4.1.6), we see that indeed $\widetilde{R}\#B$ is a Hopf subalgebra of H. □

REMARK 4.3.2. (i) Suppose that $V \subseteq R$ is a subcoalgebra satisfying condition (2) in Lemma 4.3.1; that is, suppose that there exists a Hopf subalgebra $B \subseteq A$ such that $\rho(V) \subseteq B \otimes V$ and $B.V \subseteq V$.

Then the subalgebra $k[V]$ generated by V in R is both a subalgebra and a subcoalgebra, by (4.1.4); moreover $\rho(k[V]) \subseteq B \otimes k[V]$ and $B.k[V] \subseteq k[V]$ because the multiplication of R is a comodule and module map. Therefore, Lemma 4.3.1 implies that $k[V]$ is a braided Hopf algebra over B and $k[V]\#B$ is a Hopf subalgebra of H.

(ii) Suppose that there exists a Hopf subalgebra $B \subseteq A$ such that $\rho(R) \subseteq B \otimes R$. Then Lemma 4.3.1 applies with $\widetilde{R} = R$, and we find that $R\#B$ is a Hopf subalgebra of H.

LEMMA 4.3.3. *Suppose that $V \subseteq R$ is an irreducible left coideal of H. Then we have*
(i) V is not irreducible as an A-subcomodule of R.
(ii) Assume in addition that $\dim V = 2$ and $\sum_{g \in G(A)} g.V$ generates R as an algebra. Then $\rho(R) \subseteq kG(A) \otimes R$ and $R\#kG(A)$ is a Hopf subalgebra of H.

PROOF. (i) Let $\pi := \epsilon \otimes \mathrm{id} : H \to A$ be the canonical Hopf algebra projection. We have $\rho(V) \subseteq A \otimes V$. Suppose on the contrary that V is an irreducible A-subcomodule of R.

Then $\rho(V) \subseteq C_0 \otimes V$, where $C_0 \subseteq A$ is a simple subcoalgebra of dimension $(\dim V)^2$. Let $C \subseteq H$ be the simple subcoalgebra containing V. We have $\rho(V) = (\pi \otimes \mathrm{id})\Delta(V) \subseteq \pi(C) \otimes V$, where $\pi : H \to A$ is the projection. Thus, $C_0 \subseteq \pi(C)$, and since $\dim C = \dim C_0$, we find that $\pi|_C$ is injective. This is absurd since $\pi|_V = \epsilon|_V$ because $V \subseteq R$.

(ii) By part (i), V is not irreducible as an A-comodule and therefore $\rho(V) \subseteq kG(A) \otimes V$. By (4.1.1), $\rho(g.V) \subseteq kG(A) \otimes g.V$, for all $g \in G(A)$.

This implies, in virtue of the assumption, that $\rho(R) \subseteq kG(A) \otimes R$. Thus, by Remark 4.3.2 (ii), R is a braided Hopf algebra over $kG(A)$ and the biproduct $R \# kG(A)$ is a Hopf subalgebra of H. □

Assume that $\dim A = p^3$, p prime, and A is not cocommutative. Then the index of $G(A)$ in A is p and $kG(A)$ in normal in A. Moreover, the irreducible A-comodules have dimension 1 or p. See [**M7**]. We thus obtain the following corollary.

COROLLARY 4.3.4. *Let $H = R \# A$ be a biproduct, where A is a non-cocommutative semisimple Hopf algebra of dimension p^3. Suppose that $V \subseteq R$ is an irreducible p-dimensional left coideal of H such that $\sum_{g \in G(A)} g.V$ generates R as an algebra.*
Then H contains a Hopf subalgebra of index p.

PROOF. By Lemma 4.3.3 (i) V is not irreducible as an A-comodule; therefore $\rho(V) \subseteq kG(A) \otimes V$. Then $\rho(R) \subseteq kG(A) \otimes R$, since by assumption the sum $\sum_{g \in G(A)} g.V$ generates R. Remark 4.3.2 (ii) now implies that $R \# kG(A)$ is a Hopf subalgebra of H. □

4.4. Biproducts over finite groups

Assume that $A = kG$ is the group algebra of a finite group G. Thus, R is a G-graded algebra

$$R = \bigoplus_{g \in G} R_g, \quad R_g = \{r \in R : \rho(a) = g \otimes a\},$$

such that $\Delta_R(R_g) \subseteq \bigoplus_{st=g} R_s \otimes R_t$. The action of G is by algebra and coalgebra automorphisms, and we have

(4.4.1) $$h.R_g = R_{hgh^{-1}},$$

for all $g, h \in G$. The braiding $\tau_{R,R} : R \otimes R \to R \otimes R$ is given by

(4.4.2) $$\tau_{R,R}(a \otimes b) = (g.b) \otimes a, \quad \forall a \in R_g, \ b \in R.$$

Let Supp $R \subseteq G$ denote the set of elements $g \in G$ such that $R_g \neq 0$. Let also G_R denote the subgroup of G generated by Supp R.

LEMMA 4.4.3. *G_R is a normal subgroup of G. Moreover, R is a Yetter-Drinfeld Hopf algebra over G_R with respect to the coaction ρ and the restricted action of G_R on R, and the biproduct $R \# kG_R$ is a normal Hopf subalgebra in $R \# kG$.*

PROOF. It follows from (4.4.1) that G_R is normal in G. By Remark 4.3.2 (ii), $R \# kG_R$ is a Hopf subalgebra of $R \# kG$. It is normal thanks to (4.1.6). See [**Mo2**]. □

The following lemma will help to find, in certain cases, normal Hopf subalgebras of Hopf algebras obtained as biproducts.

LEMMA 4.4.4. *(i) Suppose that G contains a normal subgroup N such that N acts trivially on R. Then the group algebra kN is a normal Hopf subalgebra in H.*

(ii) Assume that R is cocommutative and let $n = \dim R$. If $|G|$ does not divide $(n-1)!$ then there exists a subgroup $1 \neq N$ of G such that the group algebra kN is normal in H.

PROOF. (i) Since N acts trivially on R, then $ha = ah$ in H, for all $h \in N$, $a \in R$. On the other hand, the Yetter-Drinfeld condition (4.4.1) implies that $N \subseteq Z_G(G_R)$.

Note the following consequence of (4.1.6):

$$(4.4.5) \quad \mathrm{ad}_{a\#g}(b\#h) = \left(a^1 \# a^2_{-2}\right) \left(g.b \# ghg^{-1} \mathcal{S}(a^2_{-1})\right) \left(\mathcal{S}_R(a^2_0)\#1\right).$$

In particular, for all $h \in N$, we have

$$\mathrm{ad}_{a\#g}(h) = a^1 \left(a^2_{-2}(ghg^{-1})S(a^2_{-1})\right) S_R(a^2_0)$$
$$= a^1(ghg^{-1})S_R(a^2) = \epsilon(a)ghg^{-1} \in kN;$$

the second equality because $ghg^{-1} \in N \subseteq Z_G(G_R)$. This proves (i).

(ii) Since G acts on R by coalgebra automorphisms fixing $1 \in G(R)$, then G permutes the set $X = G(R)\setminus\{1\}$. This corresponds to a group homomorphism $f : G \to \mathcal{S}(X)$, where $\mathcal{S}(X)$ is the group of all permutations of X. Let N be the kernel of f. By assumption $N \neq 1$ and N acts trivially on R. Then the claim follows from part (i). □

PROPOSITION 4.4.6. *Suppose that $\dim R = 3$ or 4. Then either H or H^* contains a proper normal cocommutative Hopf subalgebra.*

PROOF. In this case, R is necessarily cocommutative. By Lemma 4.4.4, we may assume that $|G| = 2$ if $\dim R = 3$ and $|G| = 6, 3$ or 2 if $\dim R = 4$. In view of [**M4**] and [**F**], it remains to consider the case where $|G| = 6$, $\dim R = 4$. In this case, $X = G(R)\setminus\{1\}$ has three elements and we may assume that $G \simeq \mathcal{S}(X)$ is not abelian; so that $\dim H = 24$ and $|G(H)|$ is divisible by 6. We may then also assume that H is not trivial and $|G(H)| \neq 12$. Therefore $G = G(H)$ and H is of type $(1, 6; 3, 2)$ as a coalgebra. The proposition will follow from the following claim.

Claim. The group G is abelian.

Proof of the Claim. We have that $[H : G] = 4$ and $\dim R(H^*) = 8$. Consider the inclusion $kG \subseteq R(H^*)$. If G is not abelian, then $R(H^*) \simeq M_2(k) \times k^{(4)}$ as an algebra, and a complete set of orthogonal primitive idempotents of kG is of the form e_0, e_1, f_0, f_1, where $\dim He_i = 4$ and $\dim Hf_j = 8$. Moreover, f_0, f_1 are not central.

The Kac-Zhu Theorem implies that $R(H^*)$ has a primitive idempotent e with $\dim He = 2$, and by [**Z2**] $D(H)$ has an irreducible character of degree 2. By Proposition 2.3.1 $|G(D(H)^*)| \neq 1$. In view of [**R**], this implies that the subgroup of the group $G(D(H)) = G(H^*) \times G(H)$ consisting of elements which are central in $D(H)$ is not trivial. We may assume that the elements of this group are of the form $\eta \otimes g$, where $g \neq 1$ if $\eta \neq \epsilon$. Thus $Z(G) \neq 1$ and the claim follows. $\square$

4.5. Cocommutative braided Hopf algebras

Suppose that R is a cocommutative coalgebra, that is, $a^1 \otimes a^2 = a^2 \otimes a^1$, for all $a \in R$; so that all irreducible R-comodules are one-dimensional.

LEMMA 4.5.1. *Let $V \subseteq H = R \# kG$ be an irreducible left coideal of dimension $\dim V > 1$ and let $\chi = \chi_{V^*}$. Then $G[\chi] \cap G \neq 1$.*

PROOF. As coalgebras, $R \simeq H/H(kG)^+$. Let $C \subseteq H$ be the simple subcoalgebra containing V. Then the (right) stabilizer of C in G is $G[\chi] \cap G$. Consider the corestriction $\overline{V}$ of V to R as in Chapter 3.

By Corollary 3.3.2, the image of C under the canonical projection $H \to H/H(kG)^+$ is isomorphic to $C/C(k(G[\chi] \cap G))^+$. Therefore, applying Proposition 3.2.3 with $A = k(G[\chi] \cap G)$, we find an isomorphism of algebras $(\text{End}^R \overline{V})^{\text{op}} \simeq k_\alpha(G[\chi] \cap G)$, where $\alpha \in Z^2(G[\chi] \cap G, k^\times)$. Since $\dim V > 1$, $\overline{V}$ cannot be irreducible. Therefore $|G[\chi] \cap G| \neq 1$ and the lemma follows. $\square$

PROPOSITION 4.5.2. *Let $|G| = q^r$, where q is a prime number and $r > 0$. Then q divides the dimension of V, for all irreducible left coideals V of H such that $\dim V > 1$.*

Assume in addition that $R \cap G(H) = 1$. Then $\dim R = 1 \mod q$.

PROOF. Let V be an irreducible left coideal of H of dimension $\dim V > 1$. It follows from Lemma 4.5.1 that $G[\chi_V] \cap G \neq 1$ and since $G[\chi_V] \cap G \subseteq G$, we find that q divides $|G[\chi_V] \cap G|$.

Since $|G[V]|$ divides the dimension of V, it follows that $q/\dim V$.

Suppose now that $R \cap G(H) = 1$. Observe that $R \cap G(H)$ coincides with the set $G(R)^{\text{co}\rho}$ of coinvariant group-like elements in R.

Since R decomposes as a direct sum of irreducible left coideals of H: $R = k1 \oplus V_1 \oplus \cdots \oplus V_m$, and by the assumption $\dim V_j > 1$, for all $j = 1, \ldots, m$, the last claim follows. $\square$

4.6. Cocommutative braided Hopf algebras over $\mathbb{Z}_p$

We begin this section by reviewing some of the results in [**So**], concerning the classification of cocommutative cosemisimple braided Hopf algebras over groups of prime order, that will be used later. We then show some applications.

The braided Hopf algebra R is called *trivial* if the braiding $\tau_{R,R} : R \otimes R \to R \otimes R$ is the canonical flip of vector spaces, *i.e.*, $\tau_{R,R}(a \otimes b) = b \otimes a$, for all $a, b \in R$. See [**So**, Definition 1.1].

By [**Sb**] R is trivial if and only if R is a (usual) Hopf algebra; that is, if and only if $\Delta_R(ab) = a^1 b^1 \otimes a^2 b^2$, for all $a, b \in R$.

Let p be a prime number and let $\mathbb{Z}_p$ denote the cyclic group of order p. Let R be a braided semisimple Hopf algebra over $\mathbb{Z}_p$. Suppose that R is a cocommutative coalgebra. It is shown in [**So**, Proposition 7.2] that if R is nontrivial, then p divides the dimension of R. More precisely, we have the following proposition.

PROPOSITION 4.6.1. *Let $H = R\#k\mathbb{Z}_p$ such that H is not cocommutative. Suppose that R is cocommutative. Then we have:*

(i) Assume that R is trivial. Then H fits into an abelian central extension

(4.6.2) $$0 \to k\mathbb{Z}_p \to H \to R \to 1.$$

(ii) Assume that R is not trivial. Then there are exact sequences of Hopf algebras

(4.6.3) $$1 \to k^{\mathbb{Z}_p} \to H \to kF \otimes k\mathbb{Z}_p \to 1,$$

(4.6.4) $$1 \to k^{\mathbb{Z}_p} \otimes k\mathbb{Z}_p \to H \to kF \to 1,$$

for a certain group F.

PROOF. (i) Since R is cocommutative and trivial, it is isomorphic as a Hopf algebra to a group algebra $R \simeq k\Gamma$ and the action and coaction of $\mathbb{Z}_p$ on R correspond, respectively, to actions by group automorphisms $\mathbb{Z}_p \to \operatorname{Aut}\Gamma$. By [**So**, Proposition 1.11], either the action or the coaction of $\mathbb{Z}_p$ on R must be trivial. If the coaction is trivial, it follows from (4.1.6) that H is cocommutative, against the assumption. Therefore the action is trivial, and again by (4.1.6), we find that the coalgebra surjection $\pi = \operatorname{id} \otimes \epsilon : H \to R$ is indeed a Hopf algebra surjection; moreover, we have $k\mathbb{Z}_p = H^{\operatorname{co}\pi}$. Therefore, π gives rise to an abelian central extension (4.6.2).

(ii) By [**So**, Theorems 7.7 and 7.8], R is isomorphic to a crossed product $k^{\mathbb{Z}_p}\#_\sigma kF$, where F is a finite group, with 'diagonal' action and coaction of $k\mathbb{Z}_p$. These braided Hopf algebras fit into the construction described in Section 3 of [**AN3**]. Therefore, by [**AN3**, Proposition 3.7], there are exact sequences of Hopf algebras (4.6.3) and (4.6.4). □

As an application, we shall give in the following theorem a classification result for certain semisimple Hopf algebras. A stronger result (without the assumption on $|G(H^*)|$) is proved in [**IK**, Corollary IX.9] in the context of Kac algebras.

THEOREM 4.6.5. *Suppose that H is of type $(1, 2; 2, n)$ as a coalgebra. Assume in addition that $|G(H^*)|$ is even. Then H is commutative.*

PROOF. We have $\dim H = 2(2n+1)$. In particular, since 4 does not divide $\dim H$, it follows from Theorem 2.2.1 that every irreducible character of degree 2 is stable under left multiplication by $G(H)$. Therefore, by Remark 3.2.7, $H/H(kG(H))^+$ is a cocommutative coalgebra. By assumption, $G(H^*)$ contains a subgroup $\Gamma \simeq \mathbb{Z}_2$, and the projection $q : H \to k^\Gamma$ restricts injectively to $kG(H) \simeq k\mathbb{Z}_2$, because $\dim H^{\mathrm{co}\, q} = 2n + 1$ is odd. Hence $H := R \# k\mathbb{Z}_2$, where $R \simeq H/H(kG(H))^+$ is a cocommutative Yetter-Drinfeld Hopf algebra of odd dimension. By Proposition 4.6.1, since H is not cocommutative, H fits into an exact sequence $1 \to k\mathbb{Z}_2 \to H \to R \to 1$, where R is a cocommutative Hopf algebra. Hence $R = kF$, where F is a group of order $2n + 1$.

CLAIM 4.6.6. *The group F is abelian.*

PROOF. By dualizing, we get an exact sequence $1 \to k^F \to H^* \to k\mathbb{Z}_2 \to 1$. Therefore, as an algebra, H^* is isomorphic to a crossed product $H^* \simeq k^F \#_\tau k\mathbb{Z}_2$, corresponding to an action $\rightharpoonup: k\mathbb{Z}_2 \times k^F \to k^F$ and a 2-cocycle $\tau : \mathbb{Z}_2 \times \mathbb{Z}_2 \to k^F$. By [**N**], we may assume that $\tau = 1$.

The action is in this case of the form $g \rightharpoonup \delta_x = \delta_{g \triangleright x}$, for all $g \in \mathbb{Z}_2$, $x \in F$, where $\triangleright : \mathbb{Z}_2 \times F \to F$ is an action by group automorphisms (because the extension is cocentral, see [**N**]). By Clifford theory, the irreducible H^*-modules are exactly of the form $V_{x,U} = \mathrm{Ind}_{k^F \# kG_x}^{H^*} kx \otimes U$, where $x \in F$ runs over a system of representatives of the orbits of $\mathbb{Z}_2$ in F, G_x is the stabilizer of x in $\mathbb{Z}_2$ and U runs over all irreducible nonisomorphic kG_x-modules. Note that $\dim V_{x,U} = [\mathbb{Z}_2 : G_x]$. The assumption on the coalgebra structure of H implies that the action $\triangleright : \mathbb{Z}_2 \times F \to F$ has exactly one fixed point $x = 1$.

Let $\phi \in \mathrm{Aut}\, F$, $\phi(x) = g \triangleright x$, where $1 \neq g \in \mathbb{Z}_2$. Then g is an automorphism of order 2, with exactly one fixed point $x = 1$. This implies that F is abelian and moreover $g \triangleright x = x^{-1}$, for all $x \in F$. □

The proof of the theorem can be now concluded as follows: we have a Hopf algebra inclusion $k^F \subseteq H^*$. Since the group F is abelian, $k^F = k\widehat{F}$ is cocommutative, and thus $2n+1$ divides the order of $G(H^*)$.

Since by assumption the order of $G(H^*)$ is even, then $|G(H^*)| = \dim H$ and $H^* = kG(H^*)$ is cocommutative. One could alternatively use the more general statement on abelian exact sequences that we prove in Lemma 4.6.7 below. □

LEMMA 4.6.7. *Suppose that H fits into an exact sequence*

$$1 \to k^{\mathbb{Z}_2} \to H \to kF \to 1,$$

where F is an abelian group of odd order. Assume in addition that H is of type $(1, 2; 2, n)$ as a coalgebra. Then H is commutative.

PROOF. Note that the exact sequence is necessarily central, by the dual version of Corollary 1.4.3. Hence, in the associated matched pair $(\mathbb{Z}_2, F)$, the action $\triangleleft : \mathbb{Z}_2 \times F \to \mathbb{Z}_2$ is trivial and the action $\triangleright : \mathbb{Z}_2 \times F \to F$ is by group automorphisms.

As an algebra, $H \simeq k^{\mathbb{Z}_2} \#_\sigma kF$ is a crossed product for the trivial action corresponding to $\triangleleft$ and for some 2-cocycle $\sigma : F \times F \to k^{\mathbb{Z}_2}$, and as a coalgebra $H \simeq k^F \#_\tau k\mathbb{Z}_2$ with respect to the action of $\mathbb{Z}_2$ on k^F corresponding to $\triangleright$ and the *trivial* cocycle τ; see [**N**].

To establish the lemma, it will be enough to show that the cocycle σ is a coboundary, that is, it is symmetric.

Let g be the generator of $\mathbb{Z}_2$. The assumption on the coalgebra structure of H implies, as in the proof of Theorem 4.6.5, that g acts on F by $g \triangleright x = x^{-1}$ for all $x \in F$. Write

$$\sigma(x, y) = \sigma_1(x, y)\delta_1 + \sigma_g(x, y)\delta_g, \qquad x, y \in F,$$

where $\delta_1, \delta_g \in k^{\mathbb{Z}_2}$ are the canonical idempotents. The normalized cocycle condition for σ implies that $\sigma_g : F \times F \to k^\times$ is a 2-cocycle and $\sigma_1 = 1$ [**M**]. Moreover, the compatibility condition in [**M**, (4.8)] between σ and the trivial cocycle τ implies that

$$\sigma_g(x, y)\sigma_g(x^{-1}, y^{-1}) = 1, \qquad \forall x, y \in F.$$

So that $\sigma_g(x^{-1}, y^{-1}) = \sigma_g(x, y)^{-1}$, for all $x, y \in F$. But the 2-cocycle $\beta : F \times F \to k^\times$, $\beta(x, y) := \sigma_g(x^{-1}, y^{-1})$ is cohomologous to σ_g, because of the injectivity of the map

$$\phi : H^2(F, k^\times) \to \text{Hom}(\Lambda^2(F), k^\times), \qquad \phi([\alpha])(x, y) := \alpha(x, y)\alpha(y, x)^{-1}.$$

Hence, in $H^2(F, k^\times)$, we have $[\sigma_g] = [\beta] = [\sigma_g]^{-1}$, implying that $[\sigma_g]^2 = 1$, and *a fortiori* that $[\sigma_g] = 1$ since the order of F is odd. This finishes the proof of the lemma. □

The following more general form of Lemma 4.6.7 was suggested by a comment of the referee.

COROLLARY 4.6.8. *Suppose that H is of type $(1,2;2,n)$ as a coalgebra. Assume in addition that H fits into an exact sequence*

$$1 \to k^{\mathbb{Z}_2} \to H \to K \to 1,$$

where K is a Hopf algebra of odd dimension. Then K is cocommutative and H is commutative.

PROOF. Note first that K is cocommutative: this follows from Remark 3.2.7, since $K \simeq H/H(kG(H))^+$, and all simple subcoalgebras of dimension 4 are necessarily stable under left and right multiplication by $G(H)$. Thus $K \simeq kF$, for some group F of odd order. In view of Lemma 4.6.7 the corollary will follow if we show that K is also commutative. We do this as follows: consider the matched pair $(\mathbb{Z}_2, F)$ associated with H. As in the proof of Claim 4.6.6 we see that the assumption on the coalgebra structure of H implies that the action $\triangleright : \mathbb{Z}_2 \times F \to F$ is by group automorphisms and cannot have fixed points $x \neq 1$. Then the group is abelian, and K is commutative. □

4.7. Transitive actions of central subgroups

We shall assume in this section that R contains a cocommutative subcoalgebra V such that $\rho(V) \subseteq kG \otimes V$. As before, we shall use the notation G_V to indicate the subgroup of G generated by $\operatorname{Supp} V$. We have $\rho(V) \subseteq kG_V \otimes V$ and there is a basis $v_1, \ldots, v_n$ of V such that $\rho(v_i) = g_i \otimes v_i$, where $g_i \in \operatorname{Supp} V$, $\forall i = 1, \ldots, n$.

LEMMA 4.7.1. *Suppose that there exists a central subgroup S of G with the property that S permutes transitively the set $G(V)$. Then G_V is an abelian subgroup of G of order $\leq \dim V$.*

PROOF. Let $G(V) = \{a_1, \ldots, a_n\}$. Since $G(V)$ is a basis of V, we may write $\rho(a_i) = \sum_j t_{ij} \otimes a_j$, for some $t_{ij} \in kG$. As a consequence of the kG-colinearity of Δ_R we get the following relations:

(4.7.2) $$t_{ij} t_{il} = \delta_{jl} t_{ij}, \quad \forall 1 \leq i, j, l \leq n.$$

On the other hand, the coassociativity of ρ implies that, for all $p = 1, \ldots, n$, the subspace R_p spanned by the set $\{t_{pj} : 1 \leq j \leq n\}$ is a right coideal of kG; by definition, we have $kG_V = \sum_p R_p$.

The assumption on S implies that $R_p = R_j$, for all $1 \leq p, j \leq n$. Fix p, and let $e_j := t_{pj}$, $1 \leq j \leq n$. By (4.7.2) we get that kG_V, which is spanned by $\{e_j\}$, is a commutative subalgebra of kG of dimension $\leq n$. This proves the lemma. □

CHAPTER 5

Cocycle Deformations of Some Hopf Algebras

In this chapter, we describe some known examples of semisimple Hopf algebras as cocycle twists of group algebras. We also show that some others cannot be obtained in this fashion.

5.1. Lifting from abelian groups

Let G be a finite group and let $\phi \in kG \otimes kG$ be a normalized invertible 2-cocycle. The following lemma is a consequence of [**EG5**].

LEMMA 5.1.1. *The Hopf algebra $(kG)_\phi$ is cocommutative if and only if there exist a normal abelian subgroup $A \subseteq G$ and an $\operatorname{ad} G$-invariant cohomology class $[\omega] \in H^2(\widehat{A}, k^\times)$ such that*

$$(5.1.2) \qquad \phi = \sum_{x,y \in \widehat{A}} \omega(x,y) \delta_x \otimes \delta_y \in A \otimes A,$$

where, for $x \in \widehat{A}$, the element $\delta_x \in A$ is the primitive idempotent defined by $\delta_x := \frac{1}{|A|} \sum_{a \in A} x(a) a$.

PROOF. Note that $(kG)_\phi$ is cocommutative if and only if $\phi \phi_{21}^{-1}$ commutes with $\Delta(G)$. The 'if' direction follows from this observation and the invariance of $[\omega]$. The 'only if' part follows from [**EG5**, Proof of Theorem 1.3]. □

REMARK 5.1.3. Suppose that the 2-cocycle ϕ has the form (5.1.2). Then we say that ϕ is *lifted* from the (not necessarily normal) abelian subgroup A; see [**Nk**].

Suppose that ϕ is lifted from the normal abelian subgroup A. Recall the isomorphism

$$\theta : H^2(\widehat{A}, k^\times) \to \operatorname{Hom}(\Lambda^2(\widehat{A}), k^\times), \quad \theta([\sigma])(x,y) = \sigma(x,y)\sigma(y,x)^{-1}.$$

This isomorphism commutes with the $\operatorname{ad} G$-action (i.e., the action coming from the adjoint action of G on A).

Then $(kG)_\phi$ is cocommutative if and only if $[\omega]$ is $\operatorname{ad} G$-invariant in $H^2(\widehat{A}, k^\times)$.

In the rest of this chapter p and q will be distinct prime numbers.

5.2. Examples in dimension pq^2; $p = 1 \mod q$

Suppose that $p = 1 \mod q$. Let $\mathcal{A}_i$, $0 \leq i \leq q-1$ be the nontrivial semisimple Hopf algebras constructed in [**G**]. The Hopf algebras $\mathcal{A}_i$ are self-dual and we have $G(\mathcal{A}_0) \simeq \mathbb{Z}_q \times \mathbb{Z}_q$, while $G(\mathcal{A}_i) \simeq \mathbb{Z}_{q^2}$ for all $i = 1, \ldots, q-1$; see [**AN**].

Let $F = \langle a : a^p = 1 \rangle$ be the cyclic group of order p and let $\Gamma = \langle s, t : s^q = t^q = sts^{-1}t^{-1} = 1 \rangle \simeq \mathbb{Z}_q \times \mathbb{Z}_q$.

Let $1 < m, l \leq p-1$ be units modulo p such that $m^q = l^q = 1 \mod p$. Consider the semidirect product $G = F \rtimes \Gamma$ corresponding to the action by group automorphisms of Γ on F defined on generators by $s.a = a^m$ and $t.a = a^l$.

Let $\omega \in Z^2(\widehat{\Gamma}, k^\times)$ be a nontrivial 2-cocycle; in particular, the cohomology class $[\omega]$ generates $H^2(\widehat{\Gamma}, k^\times) = \text{Hom}(\Lambda^2 \widehat{\Gamma}, k^\times) \simeq \mathbb{Z}_q$. Consider the invertible normalized 2-cocycle $\phi = \sum_{x,y \in \Gamma} \omega(x,y) \delta_x \otimes \delta_y \in k\Gamma \otimes k\Gamma$.

The following proposition generalizes [**Nk**, Example 2.9]. It also follows from the results in [**M5**, Theorem 4.8] in the case where $p = 3$ and $q = 2$.

PROPOSITION 5.2.1. *(i)* $\mathcal{A}_0 \simeq (kG)_\phi$ *as Hopf algebras;*
(ii) Let $1 \leq i \leq q-1$. Then the Hopf algebras $\mathcal{A}_i$ cannot be obtained from a group algebra by twisting the comultiplication by means of a 2-cocycle.

PROOF. (i) We claim that the Hopf algebra $(kG)_\phi$ is nontrivial. Note that the twisted group algebra $k_\omega \Gamma$ is simple; so that ϕ is a minimal twist for $k\Gamma$ in the sense of [**EG4**]. Therefore we have
$$k\Gamma = \{(f \otimes \text{id})(b); f \in k^\Gamma\},$$
where $b = \phi_{21}^{-1} \phi \in k\Gamma \otimes k\Gamma$.

Suppose that $(kG)_\phi$ is cocommutative. This is equivalent to the condition $b\Delta(g) = \Delta(g)b$, for all $g \in G$, or
$$gb^1 g^{-1} \otimes gb^2 g^{-1} = b^1 \otimes b^2, \quad \forall g \in G.$$
Regard $\widehat{\Gamma}$ as a subgroup of $\widehat{G}$ by transposing the natural projection $G \to \Gamma$. For all $f \in \widehat{\Gamma}$, after applying $f \otimes \text{id}$ to both sides of the above equation, we have $g(f(b^1)b^2)g^{-1} = f(b^1)b^2$. Since $\widehat{\Gamma}$ spans k^Γ, this implies that Γ is central in G, which is a contradiction. This establishes the claim.

Since Γ is abelian and $\phi \in k\Gamma \otimes k\Gamma$, it follows $G((kG)_\phi) = \Gamma$. We know that $(kG)_\phi$ and $(kG)_\phi^*$ are of Frobenius type [**EG3**]. Therefore, by the classification results for semisimple Hopf algebras of dimension pq^2 [**N3**, Theorem 5.4.2], $(kG)_\phi \simeq \mathcal{A}_0$ as claimed.

(ii) In this case we have $G(\mathcal{A}_i) \simeq \mathbb{Z}_4$. Suppose on the contrary that $\mathcal{A}_i = (kN)_J$, for some group N and 2-cocycle $J \in kN \otimes kN$. By [**EG4**] there exists a subgroup H of N such that $J \in kH \otimes kH$ is a minimal twist for kH. In particular, $|H|$ is a square and therefore $|H| = q^2$. This forces $H \simeq \mathbb{Z}_q \times \mathbb{Z}_q$ in view of the minimality of J. But then $kH \subseteq (kN)_J$ is a Hopf subalgebra of dimension q^2, necessarily isomorphic to $kG(\mathcal{A}_i)$. This is a contradiction. Thus part (ii) follows. □

5.3. Examples in dimension pq^2; $q = 1 \mod p$

Suppose that $q = 1 \mod p$. There is a family of Hopf algebras of dimension pq^2, constructed in [**N**] as a generalization of the examples in [**M3**]. This family is parametrized by the Hopf algebras $\mathcal{B}_\lambda$, $0 \leq \lambda < p - 1$, where $\lambda\lambda' \neq 1 \mod p$.

We have $G(\mathcal{B}_\lambda) \simeq \mathbb{Z}_q \times \mathbb{Z}_q$ for all λ, $|G(\mathcal{B}_\lambda^*)| = p$ for $\lambda > 0$, and $G(\mathcal{B}_0^*) \simeq \mathbb{Z}_{pq}$.

Let F and Γ be as before. Let $G = G_\lambda$ be the semidirect product $\Gamma \rtimes F$ with respect to the action of F on Γ given by $a.s = s^n$, $a.t = t^{n^\lambda}$, where $0 < n \leq q - 1$ is such that $n^p = 1 \mod q$, and $0 \leq \lambda < p - 1$.

Let $[\omega] \in H^2(\widehat{\Gamma}, k^\times)$ be the 2-cocycle corresponding to the skew-symmetric non-degenerate bicharacter $\Omega : \widehat{\Gamma} \times \widehat{\Gamma} \to k^\times$, $\Omega(u, v) = \det(uv)$.

We have $\Omega(a.u, a.v) = n^{\lambda+1}\Omega(u, v)$. Therefore, the bicharacter Ω, and hence also the class $[\omega]$ is *not* ad G-invariant.

Consider the 2-cocycle $\phi \in kG \otimes kG$ lifted from Γ as in Remark 5.1.3, corresponding to ω.

PROPOSITION 5.3.1. *(i)* $\mathcal{B}_\lambda \simeq (kG_\lambda)_\phi$ *as Hopf algebras;*
(ii) The Hopf algebras $\mathcal{B}_\lambda^*$ are not a cocycle deformation of any finite group.

In the case where $p = 2$, part (i) of the proposition is contained in [**M10**, Proposition 3.2].

PROOF. Part (i) follows from the above discussion, in view of the classification results for semisimple Hopf algebras of dimension pq^2 and the description in [**N**, 1.4].

Part (ii) follows from an argument similar to the one used to prove part (ii) of Proposition 5.2.1, since $\mathcal{B}_\lambda^*$ contains no Hopf subalgebra of square dimension. □

5.4. Normal Hopf subalgebras in cocycle twists

Let H be a semisimple Hopf algebra and let $\phi \in (H \otimes H)^\times$ be a 2-cocycle.

LEMMA 5.4.1. *Let $B \subseteq H$ be a Hopf subalgebra of H. Then B is a Hopf subalgebra of H_ϕ if and only if $\phi(B \otimes B)\phi^{-1} = B \otimes B$. In this case, B is normal in H if and only if it is normal in H_ϕ.*

Note that the Hopf algebra structure on B as a Hopf subalgebra of H_ϕ is not *a priori* that of a 2-cocycle twisting of B, since we may not have $\phi \in B \otimes B$.

PROOF. The first part of the lemma follows easily. As for the second part, recall that B is normal in H if and only if $HB^+ = B^+H$, which depends only on the multiplication and the counit of H. □

The following corollary is an easy consequence of the lemma.

COROLLARY 5.4.2. *Suppose that $B \subseteq H$ is a central Hopf subalgebra. Then B is a central Hopf subalgebra of H_ϕ.* □

In particular, let G be a finite group and let $\phi \in kG \otimes kG$ be a two cocycle. Then if $Z(G) \neq 1$, $(kG)_\phi$ contains a nontrivial central group-like element.

The remaining chapters will be devoted, respectively, to the proof of our main results on semisimple Hopf algebras of dimension 24, 30, 36, 40, 42, 48, 54 and 56.

CHAPTER 6

Dimension 24

6.1. Possible (co)-algebra structures

Let H be a nontrivial semisimple Hopf algebra of dimension 24.

LEMMA 6.1.1. *The group $G(H^*)$ is of order 2, 3, 4, 6, 8 or 12. As an algebra, H is of one of the following types:*

- $(1, 2; 2, 1; 3, 2)$,
- $(1, 3; 2, 3; 3, 1)$,
- $(1, 4; 2, 5)$,
- $(1, 4; 2, 1; 4, 1)$,
- $(1, 6; 3, 2)$,
- $(1, 8; 2, 4)$,
- $(1, 8; 4, 1)$,
- $(1, 12; 2, 3)$.

We shall prove later that the type $(1, 4; 2, 1; 4, 1)$ is impossible, that is, there exists no semisimple Hopf algebra with this (co)algebra structure; see Lemma 6.1.7 below. The analogous result for Kac algebras follows from [**IK**, Proposition XIV.37], since there is no group G for which kG has this algebra structure.

PROOF. The proof follows from counting arguments using 1.1. □

REMARK 6.1.2. (i) If H is of type $(1, 12; 2, 3)$ as an algebra, then H is not simple by Corollary 1.4.3.

(ii) Suppose that H is of type $(1, 2; 2, 1; 3, 2)$ as an algebra. Then H has a unique irreducible character χ of degree 2, which necessarily satisfies (2.1.2) with $G(H^*) = \{\epsilon, g\}$. Therefore there is a quotient Hopf algebra $H \to k\mathbb{S}_3$, where $\mathbb{S}_3$ is the only non-abelian group of order 6.

(iii) Suppose that H is of type $(1, 4; 2, 5)$ or $(1, 4; 2, 1; 4, 1)$ as a coalgebra. Then, by Proposition 2.1.3, H contains a non-cocommutative Hopf subalgebra of dimension 8.

LEMMA 6.1.3. *Suppose that H is of type $(1, 3; 2, 3; 3, 1)$ as a coalgebra. Then $G(H^*) \cap Z(H^*) \neq 1$.*

PROOF. It follows from Remark 2.2.2 (i) that H has a Hopf subalgebra K of dimension 12. Since the index of K in H is 2, then $G(H^*) \cap Z(H^*) \neq 1$. $\square$

LEMMA 6.1.4. *Assume that H is of type $(1,6;3,2)$ as a coalgebra. Then H is not simple.*

PROOF. Suppose on the contrary that H is simple. Then H must be of type $(1,2;2,1;3,2)$ or $(1,6;3,2)$ as an algebra. Otherwise, by Lemma 6.1.1, Remark 6.1.2 and Lemma 6.1.3, there is a Hopf algebra quotient $H \to \overline{H}$, where $\dim \overline{H} = 8$. Therefore $\dim H^{\operatorname{co} \overline{H}} = 3$. Decomposing $H^{\operatorname{co} \overline{H}}$ as a direct sum of irreducible left coideals of H (of dimension 1 or 3), we find that $H^{\operatorname{co} \overline{H}} \subseteq kG(H)$ and thus $H^{\operatorname{co} \overline{H}} = kT \simeq k^T$ is a Hopf subalgebra, where T is the unique subgroup of $G(H)$ of order 3; this implies that H is not simple.

Therefore, there is a quotient Hopf algebra $\pi : H \to A$, where A is of dimension 6. We have $\dim H^{\operatorname{co} \pi} = 4$, implying that $H^{\operatorname{co} \pi} = k1 \oplus V$, where V is an irreducible left coideal of dimension 3. It follows that the restriction $\pi|_{kG(H)} : kG(H) \to A$ is an isomorphism. In particular, $A \simeq kG$ and $H \simeq R \# kG$ is a biproduct, where $G = G(H)$. The lemma follows from Proposition 4.4.6. $\square$

LEMMA 6.1.5. *Suppose that H is simple. If H has a Hopf subalgebra of dimension 8, then H is of type $(1,2;2,1;3,2)$ as an algebra.*

PROOF. By assumption, there is a Hopf algebra quotient $H^* \to B$, where $\dim B = 8$; so that $\dim(H^*)^{\operatorname{co} B} = 3$ and thus $(H^*)^{\operatorname{co} B} = k1 \oplus V$ as a left coideal of H^*, where V is an irreducible left coideal of dimension 2.

Suppose that H is not of the claimed type as an algebra. In view of Lemma 6.1.1 and the previous lemmas, there is a Hopf subalgebra $A \subseteq H^*$, with $\dim A = 8$. We have $A \cap (H^*)^{\operatorname{co} B} = k1$, unless $V \subseteq A$. However, the last possibility implies that $(H^*)^{\operatorname{co} B} \subseteq A$, which contradicts Lemma 1.3.4. Therefore, $H^* = R \# A$ is a biproduct, where R is a 3-dimensional Yetter-Drinfeld Hopf algebra over A.

By Corollary 4.3.4 and Proposition 4.4.6, H is not simple in this case. Thus H is of type $(1,2;2,1;3,2)$, as claimed. $\square$

COROLLARY 6.1.6. *Suppose that H is of coalgebra type $(1,8;2,4)$ or $(1,8;4,1)$. Then H is not simple.*

PROOF. In both cases we have $|G(H)| = 8$. By Lemma 6.1.5 and Remark 6.1.2 (ii) we may assume that there is a Hopf algebra quotient $q : H \to B$, where $\dim B = 6$. Since $\dim H^{\operatorname{co} B} = 4$, then $\dim H^{\operatorname{co} B} \cap kG(H) = 2$ or 4. The last possibility implies that H is not simple, and the first one implies that $\dim q(kG(H)) = 4$ which contradicts [**NZ**]. This proves the corollary. □

LEMMA 6.1.7. *There exists no semisimple Hopf algebra with coalgebra type* $(1, 4; 2, 1; 4, 1)$.

PROOF. Suppose on the contrary that H is of type $(1, 4; 2, 1; 4, 1)$ as a coalgebra. The set of irreducible characters of H^* consists of $G(H)$, one irreducible character λ of degree 2 and one irreducible character ψ of degree 4. Then $G(H)$ and λ span a standard subalgebra of $R(H^*)$, which corresponds to the Hopf subalgebra $B \subseteq H$, with $\dim B = 8$. By [**M4**], B is isomorphic to H_8 or to k^G, where G is either the dihedral or the quaternionic group of order 8.

It is not hard to see that the fusion rules in $R(H^*)$ are determined by $\lambda^2 = \sum_{g \in G(H)} g$, $\psi^2 = \sum_{g \in G(H)} g + 2\lambda + 2\psi$; in particular, we have $g\psi = \psi = \psi g$, for all $g \in G(H)$, and $\lambda\psi = 2\psi = \psi\lambda$.

CLAIM 6.1.8. *B is commutative.*

PROOF. Let $C \subseteq H$ be the 16-dimensional simple subcoalgebra corresponding to ψ. By Remark 3.2.2 the fusion rules for $\widehat{H^*}$ imply that $BC = C = CB$. Let $\overline{C} = C/CB^+$; so that $\overline{C}$ is a cocommutative coalgebra of dimension 2.

By Corollary 3.2.5 there is a bijective correspondence between simple $\overline{C}$-comodules and simple B_α-modules, for some invertible normalized 2-cocycle $\alpha : B \otimes B \to k$.

Suppose that B is not commutative; so that $B \simeq H_8$ is the unique nontrivial semisimple Hopf algebra of dimension 8. It follows from [**M5**, Theorem 4.8 (1)] that any Galois object of B is trivial; so that $B_\alpha \simeq B$ as B-comodule algebras. This is impossible since B has more than 2 non-isomorphic simple modules. Hence B must be commutative. □

We have $B \simeq k^G$, where G is not abelian of order 8. If B is normal in H, then H fits into an abelian extension $1 \to k^G \to H \to kF \to 1$, where F is the cyclic group of order 3.

Consider the associated matched pair $\triangleright : G \times F \to F$, $\triangleleft : G \times F \to G$; see [**M**]. There exists a subgroup G_0 of G, with $|G_0| = 4$, which acts trivially on F. The compatibility conditions between $\triangleright$ and $\triangleleft$ imply that G_0 is stable under the action of F. This implies, in view of formulas (4.2) and (4.5) in [**M**], that the subspace $A_0 = k^F \otimes kG_0$ is a Hopf subalgebra of H^* of dimension 12, necessarily normal.

We have $H \simeq k^G \#_\sigma kF$ is a crossed product, and the irreducible H-modules are of dimension 1 or 3. We may assume H is not commutative, since otherwise $H^* = k(F \bowtie G)$ would have no irreducible modules of dimension 4. By Lemma 6.1.1, H^* is of type $(1, 6; 3, 2)$ as a coalgebra. Hence A_0 is commutative, and H^* fits into an abelian extension $1 \to A_0 \to H^* \to k\mathbb{Z}_2 \to 1$. Then the irreducible H-comodules are of dimension 1 or 2. In particular, H^* cannot have irreducible modules of dimension 4. This shows that B is not normal in H.

CLAIM 6.1.9. *H is of type $(1, 2; 2, 1; 3, 2)$ as an algebra.*

PROOF. Since B is not normal in H, then $(H^*)^{\operatorname{co} B^*} = k1 \oplus V$, where V is an irreducible left coideal of dimension 2. Suppose H is not of type $(1, 2; 2, 1; 3, 2)$. Combining Theorem 2.5.1 with Lemma 6.1.1, we see that the possible coalgebra types for H^* are $(1, 4; 2, 5)$ and $(1, 8; 2, 4)$.

By Remark 6.1.2, H^* contains a Hopf subalgebra K of dimension 8. By Lemma 1.3.4, $K \cap (H^*)^{\operatorname{co} B^*} = k1$. Hence $K \simeq B^*$ is cocommutative and H^* is a biproduct $H^* \simeq R \# K$, where $\dim R = 3$.

By Lemma 4.4.4, there exists a normal subgroup $1 \neq N \subseteq G(K)$ such that N commutes with R. Since $|G(K)| = 8$, $N \subseteq Z(G(K))$; so that N commutes also with K. Therefore N is central in H^*, and H^* has a central group-like element of order 2. This is a contradiction, since H has no Hopf subalgebra of dimension 12. Hence H is of type $(1, 2; 2, 1; 3, 2)$ as an algebra, as claimed. □

The irreducible characters of degrees 1 and 2 belong to a commutative Hopf subalgebra $A \subseteq H^*$.

CLAIM 6.1.10. *A is normal in H^*.*

PROOF. Consider the projection $q : H^* \to k^{G(H)}$. Then we have $\dim(H^*)^{\operatorname{co} q} = 6$. Suppose that A is not normal. Then q is not normal, and $G(H^*) \cap (H^*)^{\operatorname{co} q} = 1$. Let $1 \neq g \in G(H^*)$. Thus there exists $s \in G(H)$ such that $\langle g, s \rangle \neq 1$. Therefore, we must have $H^{\operatorname{co} A^*} = k1 \oplus kt \oplus V$, where V is an irreducible left coideal of dimension 2. In particular, $H^{\operatorname{co} A^*} = B^{\operatorname{co} A^*} \subseteq B$.

Since $[H : B] = 3$, it follows from Lemma 2.3.2 and Proposition 2.3.1 that $G(D(H)^*) \neq 1$; hence H has a one dimensional Yetter-Drinfeld module.

Note that $G(H^*) \cap Z(H^*) = 1$ and $G(H) \cap Z(H) = 1$, since neither H nor H^* contain Hopf subalgebras of dimension 12.

By Lemma 1.6.1, a one dimensional Yetter-Drinfeld module of H is of the form $V_{a,g}$, for some $1 \neq a \in G(H)$. Consider the projection $q : H \to k^{\langle g \rangle}$. Since B is commutative, $a^{-1}ha = h$, for all $h \in B^{\operatorname{co} q}$. By Theorem 1.6.4, $B^{\operatorname{co} q}$ is a Hopf subalgebra of B, which is a contradiction. □

The above claim implies that H^* fits into the abelian extension
$$1 \to k^G \to H^* \to kF \to 1,$$
where $G \simeq \mathbb{S}_3$ and $F \simeq \mathbb{Z}_2 \times \mathbb{Z}_2$.

Then H^* is isomorphic to a crossed product $k^G \#_\sigma kF$ as an algebra. Using Clifford theory, we know that the isomorphism classes of irreducible modules of H^* are exactly the classes of the modules
$$V_{x,\rho} := \operatorname{Ind}_{k^G \# k_\alpha F_x}^{H^*} x \otimes \rho,$$
where x runs over a system of representatives of the action of F in G, $F_x \subseteq F$ is the stabilizer of x, and ρ is an irreducible representation of a twisted group algebra $k_\alpha F_x$.

We have $\dim V_{x,\rho} = \dim \rho [F : F_x]$. The assumption on the coalgebra structure of H implies that $\dim V_{x,\rho} = 4$, for some $x \in G$. This implies that $[F : F_x] = 4$ (because we cannot have $[F : F_x] = \dim \rho = 2$). Hence, the orbit of x has 4 elements, and there must exist $1 \neq y \in G$ such that $F_y = F$. This implies that $|G(H)| = 8$ and gives a contradiction. This contradiction shows that H cannot have this coalgebra structure and finishes the proof of the lemma. □

LEMMA 6.1.11. *Assume that H is of type $(1,2; 2,1; 3,2)$ as a coalgebra. Then H is not simple.*

PROOF. Suppose on the contrary that H is simple. By previous lemmas, H^* is of type $(1,4; 2,5)$ or $(1,2; 2,1; 3,2)$ as a coalgebra.

CLAIM 6.1.12. *H^* is of type $(1,4; 2,5)$ as a coalgebra.*

PROOF. Suppose not. Then H^* is of type $(1,2; 2,1; 3,2)$.

Hence, there is a Hopf algebra quotient $q : H \to B$, where $\dim B = 6$ and B is cocommutative; in particular, $\dim H^{\operatorname{co} B} = 4$. On the other hand, H contains a unique Hopf subalgebra A of dimension 6 which is not cocommutative and such that $G(H) \subseteq A$.

As a left coideal of H, $H^{\operatorname{co} B} = kG(H) \oplus V$, where V is an irreducible left coideal of dimension 2, or else $H^{\operatorname{co} B} = k1 \oplus W$, where W is an irreducible left coideal of dimension 3.

The first possibility implies $H^{\operatorname{co} B} \subseteq A$, which contradicts Lemma 1.3.4. The second possibility implies that $A \cap H^{\operatorname{co} B} = k1$ and therefore the restriction $q|_A : A \to B$ is an isomorphism. This is impossible, since A is not cocommutative. Hence the claim follows. □

By Remark 6.1.2 (iii), there is a Hopf subalgebra $B \subseteq H^*$ of coalgebra type $(1,4; 2,1)$.

Since $[H^*:B] = 3$, it follows from Lemma 2.3.2 and Proposition 2.3.1, $G(D(H)^*) \neq 1$. By Lemma 1.6.1, this is the form $V_{g,\eta}$, for some $1 \neq g \in G(H)$ and $1 \neq \eta \in G(H^*)$.

Consider the projection $q: H \to k^{\langle \eta \rangle}$. Since A is commutative, $g^{-1}ag = a$, for all $a \in A^{\text{co}\, q}$. By Theorem 1.6.4, $A^{\text{co}\, q}$ is a Hopf subalgebra of A. This implies that $A^{\text{co}\, q} = A \subseteq H^{\text{co}\, q}$. In particular, $\eta|_A = \epsilon$, and this η is the only group-like element of H^* with this property: otherwise, $A = H^{\text{co}\, G(H^*)}$ and H is not simple.

Necessarily, $(H^*)^{\text{co}\, A^*} = k1 \oplus ks \oplus U$, where U is an irreducible left coideal of B of dimension 2. It follows from Corollary 3.5.2, that $s \in G(B) \cap Z(B)$. Also, $s|_A = \epsilon$; therefore $s = \eta$.

On the other hand, $(H^*)^{\text{co}\, A^*} \subseteq (H^*)^{\text{co}\, G(H)}$. Since B cannot be contained in $(H^*)^{\text{co}\, G(H)}$ by [**NZ**], we see that $B^{\text{co}\, G(H)} = B^{\text{co}\, A^*} = k1 \oplus ks \oplus U$.

By Remark 1.6.2, $V_{s,g}$ is a Yetter-Drinfeld module of H^*. Since $s \in Z(B)$, applying again Theorem 1.6.4, we see that $B^{\text{co}\, G(H)}$ is a Hopf subalgebra of B. This is a contradiction, because of the decomposition of $B^{\text{co}\, G(H)}$ as a left coideal of H^*. This contradiction finishes the proof of the Lemma. □

6.2. Upper and lower semisolvability

Our aim in this section is to prove that semisimple Hopf algebras of dimension 24 are *both* upper and lower semisolvable. We first need the following lemma:

LEMMA 6.2.1. *Suppose that H is not simple. Then H is upper and lower semisolvable.*

PROOF. We shall show that H is lower semisolvable, and the other statement will follow by duality. Note that every semisimple Hopf algebra of dimension less than 24 is upper and lower semisolvable; see Table 1. Therefore if H has a normal quotient $H \to \overline{H}$, where $\overline{H}$ is commutative or cocommutative, then H is lower semisolvable. In particular, we may assume that $G(H^*) \cap Z(H^*) = 1$. This, combined with Lemma 6.1.3 and Remark 6.1.2 (i), allows us to suppose that H is neither of type $(1, 3; 2, 3; 3, 1)$ nor $(1, 12; 2, 3)$ as a coalgebra.

By assumption, H fits into an extension

$$1 \longrightarrow A \longrightarrow H \xrightarrow{\pi} \overline{H} \longrightarrow 1,$$

and we may assume that $\overline{H}$ is not trivial. Thus $\dim \overline{H} = 8$ or 12, and $\dim A = 3$ or 2, respectively.

In the first case, we may suppose that H is of type $(1, 6; 3, 2)$ as a coalgebra, by Lemma 6.1.1. In particular, $G(H)$ has a unique (normal) subgroup G of order 3, which must coincide with the stabilizer of all simple subcoalgebras of dimension 9, such that $A = kG$; Remark 3.2.7 now implies that the quotient $\overline{H} = H/H(kG)^+$ is cocommutative, and H is semisolvable.

Suppose finally that $\dim \overline{H} = 12$ and $\dim A = 2$. We observe that if $|G(H^*)| = 12$, then any subgroup of order 3 must be contained in $(H^*)^{\mathrm{co}\, A^*} \simeq \overline{H}^*$; thus $3/|G(\overline{H}^*)|$ and $\overline{H}^*$ is commutative or cocommutative. Hence, we may assume that $|G(H^*)| \neq 12$.

If $\overline{H}$ is not trivial, then its simple subcoalgebras are of dimension 1 or 4 [**F**]. Therefore, by Corollary 3.3.2, we may assume that H contains no simple subcoalgebra of dimension 9; this implies that $|G(H)| = 4$ or 8, by Lemma 6.1.1. Also, $4 = |G(\overline{H}^*)|$ divides $|G(H^*)|$, hence we must only consider the cases $|G(H^*)| = 4$ or 8.

It turns out in this case, that both H and H^* contain Hopf subalgebras of dimension 8. If $B \subseteq H$, $K \subseteq H^*$, with $\dim B = \dim K = 8$, then $B \cap H^{\mathrm{co}\, K^*} = k1$, by Lemma 1.3.4. Hence $H = R\#B$, where B is a Hopf subalgebra of dimension 8.

If B is not cocommutative, then by Corollary 4.3.4, H has a normal Hopf subalgebra of dimension 12, implying that $G(H^*) \cap Z(H^*) \neq 1$; hence we are done in this case. If $B = kG(H)$ is cocommutative, it follows from the proof of Lemma 4.4.4 (ii) that H has a normal Hopf subalgebra of dimension 4. Thus H is semisolvable in this case as well. This completes the proof of the lemma. $\square$

THEOREM 6.2.2. *Let H be a semisimple Hopf algebra of dimension 24. Then H is upper and lower semisolvable.*

PROOF. It will be enough to show that H is not simple. This follows from Remark 6.1.2 (i), Lemmas 6.1.3, 6.1.4 and Corollary 6.1.6, if H is of type $(1, 12; 2, 3)$, $(1, 3; 2, 3; 3, 1)$, $(1, 6; 3, 2)$, $(1, 8; 2, 4)$ or $(1, 8; 4, 1)$, respectively. Also, from Lemma 6.1.11, H is not simple if H is of type $(1, 2; 2, 1; 3, 2)$ as a coalgebra. Hence, we may assume that H is of type $(1, 4; 2, 5)$ as a coalgebra, and by Remark 6.1.2 (iii), H contains a Hopf subalgebra of dimension 8. Applying Lemma 6.1.5, we find that H^* is of type $(1, 2; 2, 1; 3, 2)$ as a coalgebra, and H is not simple by Lemma 6.1.11. $\square$

CHAPTER 7

Dimension 30

In this chapter H will denote a semisimple Hopf algebra of dimension 30 over k. By [**N**, Theorem 4.6], if H fits into an abelian extension

$$1 \to k^\Gamma \to H \to kF \to 1,$$

where Γ and F are finite groups, then H is trivial.

7.1. Possible (co)-algebra structures

We shall assume in this section that H is nontrivial, and reduce the possibilities for its algebra and coalgebra structures.

LEMMA 7.1.1. *The group $G(H^*)$ of group-like elements in H^* is of order* 2, 3, 5, 6 *or* 10. *As an algebra, H is of one of the following types:*

- $(1, 2; 2, 7)$,
- $(1, 3; 3, 3)$,
- $(1, 5; 5, 1)$,
- $(1, 6; 2, 6)$,
- $(1, 10; 2, 5)$.

PROOF. It follows from 1.1 that $n := |G(H^*)| \neq 15$, and if $n = 3, 5, 6$ or 10, these are the only possible algebra types for H. Suppose that $n = 1$; again by 1.1, it follows that the set of degrees of nonlinear irreducible representations of H has at least three elements and necessarily H has an irreducible module of degree 2. This is impossible by Corollary 2.2.3. Suppose finally that $n = 2$. Then, as an algebra, H must be either of type $(1, 2; 2, 7)$ or $(1, 2; 2, 3; 4, 1)$. By Theorem 2.2.1, $G[\chi] = G(H^*)$ is of order 2, for all irreducible character χ such that $\chi(1) = 2$. The second possibility implies, by Theorem 2.4.2, that H has a quotient Hopf algebra of dimension $2 + 4.3 = 14$, which contradicts [**NZ**]. This completes the proof of the lemma. □

REMARK 7.1.2. (i) Observe that if H is of type $(1, 5; 5, 1)$ as an algebra, the irreducible character of degree 5 is stable under multiplication by $G(H^*)$. Also, if H is of type $(1, 3; 3, 3)$ as an algebra, every irreducible character ψ of degree 3 is stable under multiplication by $G(H^*)$: this can be seen decomposing the product $\psi\psi^*$ into irreducibles and using the relation (1.2.3).

In the other cases, after Theorem 2.2.1, all irreducible characters χ of degree 2 have nontrivial isotropy, i.e., $G[\chi]$ is of order 2 for all such characters. Combining this observation with Corollary 3.3.2 and Remark 3.2.7 we find that, in any case, the quotient coalgebra $H^*/H^*(kG(H^*))^+$ is cocommutative.

(ii) Suppose that H is of type $(1, 10; 2, 5)$ or $(1, 6; 2, 6)$ as an algebra. Then the group $G(H^*)$ is abelian.

PROOF. By part (i), for all irreducible character χ of degree 2 we have $G[\chi] \neq 1$. The claim follows from Proposition 1.2.6. □

LEMMA 7.1.3. *Suppose that H is of type $(1, 2; 2, 7)$ as a coalgebra. Then H is commutative.*

PROOF. Note first that if $G(H) \subseteq Z(H)$, in view of Remark 7.1.2, H fits into an abelian extension $1 \to kG(H) \to H \to H/H(kG(H))^+ \to 1$, thus implying the lemma, in view of [**N**].

By Lemma 7.1.1, $|G(H^*)| \neq 1$. If $|G(H^*)|$ is even, the lemma follows from Theorem 4.6.5 and [**N**]. We may thus assume that $|G(H^*)|$ is odd.

Consider the projection $q : H \to B$, where $B = k^{G(H^*)}$. Since the dimension of B is odd and $G(H)$ is of order 2, we must have $G(H) \subseteq H^{\operatorname{co} B}$. Therefore, as a left coideal of H, $H^{\operatorname{co} B}$ decomposes in the form $H^{\operatorname{co} B} = kG(H) \oplus V_1 \oplus \cdots \oplus V_m$, where V_i is an irreducible left coideal of H of dimension 2, for all $i = 1, \ldots, m$. For $1 \leq i \leq m$, let C_i be the simple subcoalgebra of H containing V_i. By Remark 7.1.2, we have $gC_i = C_i = C_ig$, for all $g \in G(H)$; in particular, $kG(H) \subseteq k[C_i]$.

CLAIM 7.1.4. $G(H) \subseteq Z(k[C_i])$, for all $i = 1, \ldots, m$.

PROOF. Observe that V_i appears in $H^{\operatorname{co} B}$ with multiplicity at most 2, and moreover that V_i appears with multiplicity exactly 2 if and only if $C_i \subseteq H^{\operatorname{co} B}$. Consider first the case where $m(V_i, H^{\operatorname{co} B}) = 2$. Then $k[C_i] \subseteq H^{\operatorname{co} B}$, whence $\dim k[C_i] < \dim H$. Then $k[C_i]$ is commutative and the claim follows.

Suppose now that $m(V_i, H^{\operatorname{co} B}) = 1$. Let $V = V_i$. Note that gV and Vg are irreducible left coideals of H isomorphic to V, and $gV, Vg \subseteq H^{\operatorname{co} B}$. The multiplicity condition on V implies that $gV = V = Vg$, and the claim follows in this case from Corollary 3.5.2. □

Let $A = k[C_i : 1 \leq i \leq m]$. This is a Hopf subalgebra of H, and the claim implies that $kG(H) \subseteq Z(A)$. Hence, we may assume that $A \subsetneq H$. Since also $H^{\operatorname{co} B} \subseteq A$, a dimension argument implies that

$H^{\mathrm{co}\, B} = A$ and A is commutative. Thus the map q is normal, and H fits into an abelian extension $1 \to A \to H \to k^{G(H^*)} \to 1$. This implies the lemma in view of [**N**]. $\square$

LEMMA 7.1.5. *Suppose that H is of type $(1, 3; 3, 3)$ as a coalgebra. Then H is commutative.*

PROOF. By Lemma 7.1.3, H is not of type $(1, 2; 2, 7)$ as an algebra. Suppose first that $|G(H^*)|$ is divisible by 3. Then, by Lemma 4.1.9, H is isomorphic to a biproduct $H \simeq R \# kG(H)$, where R is cocommutative by Remark 7.1.2 (i). Then the lemma follows in this case by Proposition 4.6.1 and [**N**]. Hence we may assume that H is not of type $(1, 3; 3, 3)$ nor $(1, 6; 2, 6)$ as an algebra. Similarly, if H is of type $(1, 10; 2, 5)$ as an algebra, then H fits into an exact sequence $1 \to kG(H) \to H \to k^{G(H^*)} \to 1$, and $k^{G(H^*)} \simeq H/H(kG(H))^+$ is cocommutative. Hence the result follows by [**N**].

Suppose finally that H is of algebra type $(1, 5; 5, 1)$. Let C be the unique simple subcoalgebra of dimension 25 of H^*. It is clear that $H^* = k[C]$ is generated by C as an algebra. Consider the projection $q : H^* \to k^{G(H)}$; we have $\dim(H^*)^{\mathrm{co}\, q} = 10$. Since $|G(H)|$ and $|G(H^*)|$ are relatively prime, we have $G(H^*) \subseteq (H^*)^{\mathrm{co}\, q}$. By dimension restrictions, $(H^*)^{\mathrm{co}\, q} = kG(H^*) \oplus V$, where $V \subseteq C$ is an irreducible left coideal of dimension 5. Let g be a generator of $G(H^*)$; since V is the only 5-dimensional irreducible coideal contained in $(H^*)^{\mathrm{co}\, q}$, then we must have $gV = V = Vg$. By Corollary 3.5.2 $kG(H^*)$ is normal in H^*. This implies the claim in this case, in view of [**N**], since the quotient Hopf algebra $H^*/H^*(kG(H^*))^+$ is cocommutative, by Remark 7.1.2 (i). This completes the proof of the lemma. $\square$

LEMMA 7.1.6. *Suppose that H is of type $(1, 6; 2, 6)$ as an algebra. Then H is of type $(1, 6; 2, 6)$ as a coalgebra.*

PROOF. Suppose not. It follows from Lemmas 7.1.1, 7.1.3 and 7.1.5, that $G(H)$ has a subgroup T of order 5. Since $|G(H^*)| = 6$, we have a sequence of Hopf algebra maps

$$(7.1.7) \qquad kT \xrightarrow{\iota} H \xrightarrow{\pi} k^{G(H^*)}.$$

In particular, $\pi(a) = 1$, for all $a \in T$, implying that $kT \subseteq H^{\mathrm{co}\, \pi}$, and since $\dim H^{\mathrm{co}\, \pi} = [H^* : kG(H^*)] = |T|$, we have $kT = H^{\mathrm{co}\, \pi}$. Therefore (7.1.7) is an exact sequence of Hopf algebras. By Remark 7.1.2 (ii), kT and $k^{G(H^*)}$ are both commutative and cocommutative; hence the extension is abelian. This implies that H is trivial, against the assumption. This finishes the proof of the lemma. $\square$

LEMMA 7.1.8. *Suppose that H is of type $(1, 5; 5, 1)$ as a coalgebra. Then H is commutative.*

PROOF. By Lemmas 7.1.1, 7.1.3, 7.1.5 and 7.1.6, $G(H^*)$ is of order 5 or 10. Let $\Gamma \subseteq G(H^*)$ be a subgroup of order 5, and consider the sequence of Hopf algebra maps $kG(H) \xrightarrow{\iota} H \xrightarrow{\pi} k^\Gamma$. Then $\dim H^{co\,\pi} = [H^* : k\Gamma] = 6$ and $kG(H) \cap H^{co\,\pi} = k1$ by [**NZ**]. Then the composition $\pi\iota : kG(H) \to k^\Gamma$ is an isomorphism, and therefore H is isomorphic to a biproduct $R \# kG$, where $G = G(H)$ is a group of order 5 and R is a semisimple Hopf algebra of dimension 6 in the category of Yetter-Drinfeld modules over G.

In particular, $R \simeq H/H(kG(H))^+$ as a coalgebra, and thus R is cocommutative by Remark 7.1.2 (i). By Proposition 4.6.1, H fits into an abelian extension and the result follows from [**N**]. □

LEMMA 7.1.9. *Suppose that H is of type $(1, 10; 2, 5)$ as a coalgebra. Then H is commutative.*

PROOF. By previous lemmas H is, as an algebra, necessarily of type $(1, 10; 2, 5)$. In view of Lemma 4.1.9, this implies that $G(H) \simeq G(H^*)$ are abelian and H is isomorphic to a biproduct $R \# kG(H)$, where $\dim R = 3$. Thus R is cocommutative and commutative, and since the action of $G(H)$ on R is by coalgebra automorphisms, we have that the unique subgroup of order 5, Γ, of $G(H)$ acts trivially on R.

Then $\Gamma \subseteq Z(H)$, and there is an exact sequence $1 \to k\Gamma \to H \to \overline{H} \to 1$, where $\dim \overline{H} = 6$. Note that $\overline{H}$ is cocommutative; otherwise $\overline{H} \simeq k^F$, where F is the only non-abelian group of order 6. But this implies that $F = G((k^F)^*)$ is a subgroup of $G(H^*)$, which is a contradiction. The lemma follows from [**N**]. □

7.2. Classification

In this section we aim to give a proof of Theorem 2. For this, we shall consider the possible structures given by Lemma 6.1.1.

Proof of Theorem 2. By Lemmas 7.1.1, 7.1.2, 7.1.8 and 7.1.9, we may assume that H and H^* are both of type $(1, 6; 2, 6)$ as coalgebras. In view of Remark 7.1.2 (ii), the group $G(H)$ is cyclic.

Let F be the unique subgroup of order 2 of $G(H)$. By Remark 7.1.2 (i) all irreducible characters of degree 2 are stable under multiplication by F, and therefore the quotient coalgebra $H/H(kF)^+$ is cocommutative. Clearly, H is a biproduct $R \# kF$, and $R \simeq H/H(kF)^+$ is a cocommutative coalgebra. This implies that H fits into an abelian extension in view of Proposition 4.6.1, and H is trivial by [**N**]. □

CHAPTER 8

Dimension 36

8.1. Reduction of the problem

Let H be a nontrivial semisimple Hopf algebra of dimension 36 over k.

LEMMA 8.1.1. *The order of $G(H^*)$ is either 2, 3, 4, 6, 9, 12 or 18 and as an algebra H is of one of the following types:*
- $(1, 2; 2, 4; 3, 2)$,
- $(1, 3; 2, 6; 3, 1)$,
- $(1, 4; 2, 8)$,
- $(1, 4; 2, 4; 4, 1)$,
- $(1, 4; 4, 2)$,
- $(1, 6; 2, 3; 3, 2)$,
- $(1, 9; 3, 3)$,
- $(1, 12; 2, 6)$,
- $(1, 18; 3, 2)$.

By [**GN**] there is a simple Hopf algebra $\mathcal{H}$ with algebra and coalgebra type $(1, 4; 2, 4; 4, 1)$. By construction, $\mathcal{H} = (kG)_\phi$ is a twisting of the group algebra of $G = D_3 \times D_3$ with respect to a non-degenerate 2-cocycle $\phi \in \mathbb{Z}_2 \times \mathbb{Z}_2$. Moreover, $\mathcal{H}$ is the only twisting of a group of order 36 that is simple, and we have $\mathcal{H} \simeq \mathcal{H}^{*\,\mathrm{cop}}$ [**GN**, 4.3].

PROOF. It follows from 1.1 that $G(H^*) = 1$ is impossible. It follows as well that the only possibilities for the algebra types are the prescribed ones if $|G(H^*)|$ equals either 3, 4, 6, 9, 12 or 18.

Finally, if $|G(H^*)| = 2$, a counting argument gives that H must be of type $(1, 2; 2, 4; 3, 2)$ or $(1, 2; 3, 2; 4, 1)$. In the last case, H has two characters of degree 1, ϵ and g, two irreducible characters of degree 3, ψ_1 and ψ_2, and one irreducible character of degree 4, ζ. Since $G[\psi_i] = 1$, because $G(H^*)$ is of order 2, then $g\psi_1 = \psi_2$; on the other hand, $g\zeta = \zeta$ because ζ is the only irreducible character of degree 4. Counting dimensions, we obtain $\psi_i \psi_i^* = \epsilon + 2\zeta$, $i = 1, 2$. Thus $m(\psi_i, \zeta\psi_i) = m(\zeta, \psi_i \psi_i^*) = 2$, and therefore

$$\zeta\psi_1 = 2\psi_1 + r\psi_2 + s\zeta = g\zeta\psi_1 = 2\psi_2 + r\psi_1 + s\zeta,$$

implying that $r = 2$ and $\zeta\psi_1 = 2\psi_1 + 2\psi_2 = \zeta\psi_2$.

Since $*$ fixes ζ and permutes the set $\{\psi_1, \psi_2\}$, and also since $\psi_i\zeta = (\zeta\psi_i^*)^*$, we find that $m(\zeta, \psi_i\zeta) = 0$, $i = 1, 2$. But $m(\zeta, \psi_i\zeta) = m(\psi_i, \zeta\zeta^*)$, so that $\zeta\zeta^*$ decomposes in the form $\zeta\zeta^* = \epsilon + g + r\zeta$, $r \in \mathbb{Z}^+$. Taking degrees we see that this is impossible. This discards the possibility $(1, 2; 3, 2; 4, 1)$ for the algebra type of H and finishes the proof of the lemma. □

REMARK 8.1.2. (i) Suppose that H is of type $(1, 18; 3, 2)$. Then, by Corollary 1.4.3, $G(H^*) \cap Z(H^*) \neq 1$.

(ii) Suppose that H is of type $(1, 2; 2, 4; 3, 2)$ as a coalgebra. Since $|G(H)| = 2$, H contains no Hopf subalgebra of dimension 12. By Remark 2.4.3 (i), the irreducible characters of degrees 1 and 2 give rise to a Hopf subalgebra of dimension 18. Therefore, in view of Corollary 1.4.3, $G(H^*) \cap Z(H^*) \neq 1$.

(iii) Suppose that H is of type $(1, 6; 2, 3; 3, 2)$ as a coalgebra. Then for every irreducible character χ of degree 2, we must have $|G[\chi]| = 2$ (because there are only 3 such characters) and H has no simple subcoalgebra of dimension 16. Hence, also here, the irreducible characters of degrees 1 and 2 give rise to a Hopf subalgebra of dimension 18. Therefore $G(H^*) \cap Z(H^*) \neq 1$.

(iv) If H is of type $(1, 4; 4, 2)$ as a coalgebra, then H does not contain any left coideal subalgebra of dimension 3. Therefore H^* contains no Hopf subalgebra of dimension 12.

LEMMA 8.1.3. *Suppose that H is of type $(1, 3; 2, 6; 3, 1)$ as a coalgebra. Then H is not simple.*

PROOF. By Theorem 2.2.1, H has a unique Hopf subalgebra A of dimension 12, which contains $G(H)$ and the unique simple subcoalgebra of dimension 9 of H.

The classification of semisimple Hopf algebras of dimension 12 [**F**] implies that A is commutative. Therefore $\dim V \leq [H : A] = 3$ for all irreducible H-module V; see for instance Corollary 3.9 in [**AN2**]. Then, by the previous remark, we may assume that H^* is, as a coalgebra, of one of the types $(1, 3; 2, 6; 3, 1)$, $(1, 4; 2, 8)$, $(1, 9; 3, 3)$ or $(1, 12; 2, 6)$.

Suppose that H^* is of type $(1, 9; 3, 3)$ as a coalgebra. We have a Hopf algebra projection $q : H^* \to A^*$, such that $\dim H^{*\text{co }A^*} = 3$. This projection is necessarily normal, thus H is not simple in this case.

If H^* is of type $(1, 3; 2, 6; 3, 1)$ as a coalgebra, then also H^* contains a commutative Hopf subalgebra $B \simeq A$ of dimension 12. We have a Hopf algebra projection $q : H \to B^*$, for which $\dim H^{\text{co }B^*} = 3$. Then we may assume $A \cap H^{\text{co }B^*} = k1$ and the restriction $q|_A : A \to B^*$

is an isomorphism. Thus A is cocommutative, which is absurd. This contradiction discards this type.

Similarly, if H^* is of type $(1, 12; 2, 6)$ as a coalgebra, then H is a biproduct $H = R\#A$, where R is a braided Hopf algebra over A of dimension 3. Since the irreducible A-comodules are of dimension 1 or 3, we must have $\rho(R) \subseteq kG(A) \otimes R$. This implies that R is a braided Hopf algebra over $kG(A)$ and the biproduct $K = R\#kG(A)$ is a Hopf subalgebra of H. But $\dim K = 9$ and K is thus cocommutative, which contradicts $|G(H)| = 3$. Thus this type is not possible.

Finally suppose that H^* is of type $(1, 4; 2, 8)$ as a coalgebra. Consider the surjective Hopf algebra map $H^* \to A^*$, so that $\dim(H^*)^{\mathrm{co}\,A^*} = 3$ and we may assume that $(H^*)^{\mathrm{co}\,A^*} = k1 \oplus V$ as a left coideal of H^*, where V is an irreducible coideal of dimension 2. By Theorem 2.5.1 H^* contains a Hopf subalgebra B of dimension 12 such that B is not cocommutative. As in the previous paragraphs, it turns out that $A \simeq B^*$, which is a contradiction. This finishes the proof of the lemma. □

LEMMA 8.1.4. *Suppose that H is of type $(1, 9; 3, 3)$ as a coalgebra. Then H is not simple.*

PROOF. If $|G(H^*)| = 9$, then $H = R\#kG(H)$ is a biproduct and the lemma follows from Proposition 4.4.6. Thus, by Remark 8.1.2 and Lemma 8.1.3, we may assume that $|G(H^*)| = 12$ or 4. In any case, there is a quotient $H \to k\Gamma$, where Γ is a group of order 4, for which necessarily $kG(H) = H^{\mathrm{co}\,k\Gamma}$. Hence H is not simple. □

LEMMA 8.1.5. *Suppose that H is simple. Then the order of $G(H^*)$ equals either 4 or 12. Moreover, H is isomorphic to a biproduct $H \simeq R\#kG$, where G is a group of order 4 and R is a Yetter-Drinfeld Hopf algebra over G of dimension 9.*

PROOF. By Remark 8.1.2 and Lemmas 8.1.3 and 8.1.4, $|G(H^*)|$ is not 18, 9, 6, 3 or 2. Therefore, by Lemma 8.1.1, $|G(H^*)| = 4$ or 12.

Let $\Gamma \subseteq G(H^*)$ be a subgroup of order 4. Consider the quotient $p : H \to k^\Gamma$. Then $\dim H^{\mathrm{co}\,p} = 9$. Let G be a subgroup of $G(H)$ of order 4. Then, by [**NZ**], $kG \cap H^{\mathrm{co}\,p} = k1$ and thus the restriction of p gives an isomorphism $kG \to k^\Gamma$. This implies the lemma. □

LEMMA 8.1.6. *Suppose that H is of type $(1, 4; 2, 4; 4, 1)$ as a coalgebra. Then every irreducible left coideal of dimension 2 in H is contained in a Hopf subalgebra of dimension 12.*

Moreover, H contains two Hopf subalgebras B and B', which are of type $(1, 4; 2, 2)$ as coalgebras and such that
  (i) $B \cap B' = kG(H)$ and $G(H) \simeq \mathbb{Z}_2 \times \mathbb{Z}_2$;
  (ii) $B \cup B'$ generates H as an algebra.

PROOF. Since H contains no irreducible coideals of dimension 3, then for all irreducible χ such that $\deg \chi = 2$, we have $G[\chi] \neq 1$.

CLAIM 8.1.7. *The action of $G(H)$ by left multiplication on the set X_2 of irreducible characters of degree 2 has two disjoint orbits $\{\chi, \chi'\}$ and $\{\psi, \psi'\}$. In addition, we have $G[\chi] = G[\chi']$ and $G[\psi] = G[\psi']$ are distinct subgroups of order 2 of $G(H)$.*

PROOF. Note that the tensor product $\chi \chi'$ is irreducible for some irreducible characters χ and χ' of degree 2, since otherwise, there would be a Hopf subalgebra of dimension $4 + 4.4 = 20$, which is impossible. Therefore there exist $\chi, \psi \in X_2$ such that $G[\chi]$ is not contained in $G[\psi]$; c.f. 2.4.1. Thus $G[\chi]$ and $G[\psi]$ are distinct subgroups of order 2 of $G(H)$. In particular, the group $G(H)$ is isomorphic to $\mathbb{Z}_2 \times \mathbb{Z}_2$.

Since $|G[\chi]| = |G[\psi]| = 2$, then the orbits $G(H)\chi$ and $G(H)\psi$ are of order 2. Moreover, since $G[\chi'] = G[\chi]$, for all $\chi' \in G(H)\chi$ (because $G[\chi]$ is normal in $G(H)$), then $G(H)\chi$ and $G(H)\psi$ are disjoint orbits. This proves the claim. $\square$

CLAIM 8.1.8. *We have $\chi^* = \chi$, for all $\chi \in R(H^*)$.*

PROOF. We have already shown that $a^2 = 1$, for all $a \in G(H)$. If χ is irreducible of degree 2, then we have $\chi \chi^* = \sum_{a \in G[\chi]} a + \lambda$, for some $\lambda \in X_2$ such that $\lambda^* = \lambda$. Also $G[\chi] \subseteq G[\lambda]$, which implies that $\lambda \in G(H)\chi$. Hence every orbit has a self-dual element. Moreover, if $g \in G(H)$ is such that $g\lambda = \chi$, then $\chi^* = \lambda g$ and therefore $G[\chi] = G[\lambda] \subseteq G[\chi^*]$. Hence $\chi^* \in G(H)\chi$. This implies, since each orbit has two elements, that $\chi^* = \chi$. Since χ was arbitrary, the claim follows. $\square$

The claim implies the character algebra $R(H^*)$ is commutative.

As before, we have decompositions $\chi^2 = \chi \chi^* = \sum_{a \in G[\chi]} a + \lambda$, where $\lambda \in G(H)\chi$. On the other hand, if $\chi' \in G(H)\chi$, say $b\chi = \chi'$, then $\chi' \chi = b\chi^2 = b + ba + b\lambda$, and $b\lambda \in G(H)\chi$. Thus, the set $G(H) \cup G(H)\chi$ spans a standard subalgebra of $R(H^*)$ which corresponds to a Hopf subalgebra A of dimension 12 such that $R(A^*) = k(G(H) \cup G(H)\chi)$.

The same argument applies to the other orbit and concludes the proof of the lemma. $\square$

LEMMA 8.1.9. *Suppose that H is of type $(1, 4; 2, 8)$ as a coalgebra. Assume in addition that H is simple. Then H contains two Hopf subalgebras B and B', which are of type $(1, 4; 2, 2)$ as coalgebras and such that B and B' satisfy the conditions (i) and (ii) of Lemma 8.1.6.*

PROOF. First note that it is enough to show that H contains Hopf subalgebras B, B' of dimension 12 such that $B \cap B' = kG$. Indeed, in this case, a dimension argument implies that necessarily the subalgebra generated by B and B' coincides with H. On the other hand, this also implies that the group $G = G(H)$ is isomorphic to $\mathbb{Z}_2 \times \mathbb{Z}_2$: otherwise G would contain a unique subgroup of order 2, which would be central both in B and in B' ([**F**]) and a fortiori in H.

By Lemma 8.1.5 $H = R\#kG$, where $G = G(H)$ and $\dim R = 9$. Therefore R decomposes in the form $R = k1 \oplus V_1 \oplus V_2 \oplus V_3 \oplus V_4$ as a left coideal of H, where V_i is an irreducible left coideal of dimension 2, for all $i = 1, \ldots, 4$.

It follows from Corollary 4.2.2 and Proposition 4.2.1 that V_i is a subcoalgebra of R, for all i, and V_i is not isomorphic to V_j, as left coideals of H, if $i \neq j$. Also $|G[\chi_i]| = 2$, where $\chi_i = \chi_{V_i}$. In particular, $\#G\chi_i = 2$, for all i. We have $V_i\#kG = C_i \oplus C_i h$, where C_i is the simple subcoalgebra of H containing V_i and $C_i h \neq C_i$. Thus, if $V_j \simeq V_i h$ then $V_j\#kG = V_i\#kG$ and $V_j = (\mathrm{id}\otimes\epsilon)(V_j\#kG) = V_i$. Therefore χ_i, $1 \leq i \leq 4$, form a system of representatives of the orbits of the action of G on X_2 by right multiplication.

The multiplicity-one condition for the decomposition of R implies that $V_1, \ldots, V_4$ are uniquely determined, and these are the only irreducible left coideals of H contained in R. Thus the action of G on R must permute the V_i's.

CLAIM 8.1.10. *Let $1 \leq i \leq 4$. There exists a subgroup $1 \neq S$ of G such that $s.V_i = V_i$.*

PROOF. The possible decompositions of R as an H-Yetter-Drinfeld submodule of H are the following

$$k1 \oplus X_1 \oplus X_2, \quad k1 \oplus Y_1 \oplus Y_2 \oplus Y_3 \oplus Y_4, \quad k1 \oplus X_1 \oplus Y_1 \oplus Y_2, \quad k1 \oplus Y_1 \oplus Z,$$

where $\dim X_i = 4$, $\dim Y_i = 2$, $\dim Z = 6$. Decomposing the X_i's and Z as left coideals of H, and using that the action of G permutes the V_i's, the claim follows. □

Let $1 \neq S$ be a subgroup of G such that $s.V_1 = V_1$, for all $s \in S$, as in Claim 8.1.10. The action of S must permute either transitively or trivially the set $G(V_1)$.

Suppose first that the action of S on V_1 is trivial. Let $V_1' = G.V_1$. Then the action of S on V_1' is also trivial and so is the action of S on the subalgebra $k[V_1']$ generated by V_1'.

Since $k[V_1']$ is a Yetter-Drinfeld Hopf subalgebra of R, $\dim k[V_1']$ is either 3 or 9. The last possibility implies that the action of S on R is trivial. Therefore, we may assume that $\dim k[V_1'] = 3$, and then V_1 is stable under the action of G. Thus $B = k[V_1]\#kG$ is a Hopf subalgebra of H of dimension 12. Moreover, since the action of G permutes the set $\{V_1, \ldots, V_4\}$ and by the above fixes V_1, we may assume that also $gV_2g^{-1} = V_2$, for all $g \in G$. Then the Hopf subalgebras B and $B' = k[V_2]\#kG$ do the job.

Suppose finally that the action of S on V_1 is transitive. By Lemma 4.7.1 $|G_{V_1}| = 2$. Let $V'_1 := G_{V_1}.V_1$ as above; note that $G_{V'_1} = G_{V_1}$. Then, by Lemma 4.3.1, $\widetilde{B} = k[V'_1]\#kG_{V_1} = k[V'_1]\#kG_{V'_1}$ is a Hopf subalgebra of H. If $k[V'_1] = R$, then $\widetilde{B}$ is an index-2 (and therefore normal) Hopf subalgebra of H. So we may assume that $\dim k[V'_1] = 3$, implying that $g.V_1 = V_1$ for all $g \in G$. We may also assume that $g.V_2 = V_2$ for all $g \in G$. Let $B = k[V_1]\#kG$ and $B' = k[V_2]\#kG$. Then we have $\dim B = \dim B' = 12$, and we are done. □

LEMMA 8.1.11. *Suppose that H is of coalgebra type $(1,4;2,4;4,1)$ or $(1,4;2,8)$. Assume in addition that H is simple. Then $H \simeq \mathcal{H}$.*

Here $\mathcal{H}$ denotes the simple (self-dual) example in [**GN**].

PROOF. Let B and B' be the Hopf subalgebras of H as in Lemmas 8.1.6 and 8.1.9, respectively. Then neither B nor B' is cocommutative.

CLAIM 8.1.12. *We have $B \simeq B' \simeq \mathcal{A}_0$; see Chapter 5.*

PROOF. Suppose not. Since $G(H) \simeq \mathbb{Z}_2 \times \mathbb{Z}_2$, then B and B' are not isomorphic to $\mathcal{A}_1$. In virtue of the classification in dimension 12 [**F**], at least one of them, say B, is commutative. Then there exists $1 \neq g \in Z(B') \cap G(B')$. Since $G(B') = G(B) \subseteq B$, g also commutes with B. Since B and B' generate H as an algebra, this implies that g is a central group-like element in H, contradicting the assumption. □

We know from Chapter 5 that there exists a normalized 2-cocycle $\phi \in kG(H) \otimes kG(H)$ such that $B_\phi \simeq B'_\phi \simeq kG$, where G is the group in Proposition 5.2.1. Moreover, B_ϕ and B'_ϕ are Hopf subalgebras of H_ϕ and $B_\phi \cup B'_\phi$ generates H_ϕ as an algebra, because the multiplication was unchanged. It turns out that H_ϕ is cocommutative.

More precisely, $H_\phi = kG'$, where $G' = F' \rtimes \Gamma$ is a semidirect product with $|F'| = 9$ and $\Gamma = G(H)$. Since H is simple by assumption, then by the results in [**GN**, 4.3], $G \simeq D_3 \times D_3$ and $H \simeq \mathcal{H}$. □

LEMMA 8.1.13. *Suppose that H is of type $(1, 12; 2, 6)$ as a coalgebra. Then H is not simple.*

PROOF. By Lemma 8.1.5, we may assume that $|G(H^*)| = 4$ or 12. By Lemma 8.1.11, if $|G(H^*)| = 4$ we may also assume that H^* is of type $(1, 4; 4, 2)$ as a coalgebra; this is impossible in view of Remark

8.1.2 (iv). Therefore, we must consider the case $|G(H^*)| = 12$, i.e., H and H^* are both of type $(1, 12; 2, 6)$.

Consider the quotient Hopf algebra $p : H \to k^{G(H^*)}$. We have $\dim H^{\mathrm{cop}} = 3$. If $g \in H^{\mathrm{cop}}$ for some $g \in G(H)$, then by [**NZ**] g is of order 3 and $H^{\mathrm{cop}} = k\langle g \rangle$ is a normal Hopf subalgebra of H, implying that H is not simple. Thus, we may assume that $p(g) \neq 1$, for all $g \in G(H)$. This implies that the restriction $p|_{kG(H)} : kG(H) \to k^{G(H^*)}$ is an isomorphism. Therefore the groups $G(H)$ and $G(H^*)$ are isomorphic and abelian and H is a biproduct $H = R\#kG(H)$, where $\dim R = 3$. The lemma follows from Proposition 4.4.6. □

8.2. Main result

In view of the previous results, the only case it remains to consider is that where H and H^* are both of type $(1, 4; 4, 2)$ as coalgebras.

THEOREM 8.2.1. *Let H be a semisimple Hopf algebra of dimension 36. Suppose that H is simple. Then $H \simeq \mathcal{H}$.*

PROOF. By Lemma 8.1.5, H is a biproduct $H = R\#kG(H)$, where R is a left coideal subalgebra of dimension 9. Therefore, as a left coideal of H, $R = k1 \oplus V_1 \oplus V_2$, where V_i are irreducible left coideals of dimension 4, $i = 1, 2$.

CLAIM 8.2.2. *There exists a normalized invertible 2-cocycle $\phi \in kG(H) \otimes kG(H)$ such that H_ϕ is not simple.*

PROOF. We have $gV_i \simeq V_i \simeq V_ig$ as a left coideal of H, for all $g \in G(H)$; so that by Corollary 4.2.2, V_i is a subcoalgebra of R in the category of Yetter-Drinfeld modules over $G(H)$, and we have $V_i\#kG(H)$ is a simple subcoalgebra of dimension 16. It is also clear that V_i generates R as an algebra, because $k[V_1]$ is a braided Hopf subalgebra of R containing V_i; then $\dim V_i^G = 1$ by [**N3**, Lemma 4.4.1].

Write $G = G(H)$. By [**N2**, 1.3] there exist normalized 2-cocycles $\alpha_i : G \times G \to G$ such that $V_i \simeq (k_{\alpha_i}G)^*$, $i = 1, 2$, and moreover, in view of the results in [**N3**, Section 4] there exists a normalized invertible 2-cocycle $\phi \in kG \otimes kG$ such that $H_\phi = R_\phi \# kG$, and $V := V_1$ is a cocommutative subcoalgebra of R_ϕ. See [**N3**, Remark 4.4.3].

Suppose that H_ϕ is simple. Then $H_\phi \simeq H$ as algebras and as coalgebras, whence $V\#kG(H)$ is a simple subcoalgebra of H_ϕ. The claim can be established using the argument in the proof of [**N3**, Theorem 1.0.1]. We sketch this proof below for the sake of completeness.

Write $G(V) = \{x_1, \ldots, x_4\}$. By [**N3**, Lemma 2.2.1] $(\mathcal{S}_{R_\phi}V)V$ is a Yetter-Drinfeld subcoalgebra of R_ϕ, and $(\mathcal{S}_{R_\phi}V)x_i$ is a left coideal of R containing 1, for all $i = 1, \ldots, 4$, where $\mathcal{S}_{R_\phi}$ is the antipode of R_ϕ.

Since the dimensions of the irreducible comodules of a twisted dual group algebra of G divide the order of G, we have that the dimensions of the irreducible R_ϕ-comodules are either 1 or 2.

The group G acts on R_ϕ by $\leftharpoonup: R_\phi \otimes kG \to R_\phi$, $r \leftharpoonup g := (g, r_{-1})r_0$, $r \in R_\phi$, where (,) is a non-degenerate bicharacter on G. Moreover, this action restricts to an action of G on $(\mathcal{S}_{R_\phi}V)V$ by coalgebra automorphisms. Let $\omega_1, \ldots, \omega_l$ be the group-like elements of R_ϕ which belong to $(\mathcal{S}_{R_\phi}V)V$. Then the action $\leftharpoonup$ of G permutes the set $\{\omega_1, \ldots, \omega_l\}$. As in [**N3**, Claim 5.3.1], one can see that every group-like element ω in $(\mathcal{S}_{R_\phi}V)V$ belongs to $(\mathcal{S}_{R_\phi}V)x_i$ for all $1 \leq i \leq n$.

Therefore $\mathcal{S}_{R_\phi}Vx_i = k\omega_1 \oplus \cdots \oplus k\omega_l \oplus \bigoplus_j W_j$, for all i, where W_j are irreducible right coideals of R_ϕ of dimension bigger than 1. Thus l is even, and since $1 \in \{\omega_1, \ldots\}$, there exists a nontrivial group-like element in R_ϕ invariant under the action $\leftharpoonup$ of G; this is exactly the same as a coinvariant group-like element in R_ϕ. Then $G(H_\phi) \cap R_\phi$ has an element of order 3. This is a contradiction, because $G(H_\phi)$ is of order 4. The contradiction comes from the assumption that H_ϕ is simple, hence the claim follows. $\square$

Let $A \subseteq H_\phi$ be a proper normal Hopf subalgebra. Since $H_\phi \simeq H$ as algebras, H_ϕ contains no Hopf subalgebras of dimension 12, by Remark 8.1.2 (iv). Then $\dim A = 2, 3, 4, 6$ or 18. Note that $kG(H)_\phi = kG(H)$ is a Hopf subalgebra of H_ϕ; so that 4 divides the order of $G(H_\phi)$. If $\dim A = 3$, then either $|G(H_\phi)| = 12$, which contradicts Remark 8.1.2 (iv), or else H_ϕ is cocommutative. The last possibility implies $H \simeq \mathcal{H}$ by [**GN**, 4.3]; hence we may assume that $\dim A \neq 3$.

If $\dim A = 2$, then $A \subseteq kG(H_\phi) \cap Z(H_\phi)$ by Corollary 1.4.3. Thus $A \subseteq kG(H) \cap Z(H)$; see Corollary 5.4.2. Also, if $\dim A = 4$, then there is a quotient Hopf algebra $H_\phi \to H_\phi/H_\phi A^+ =: \overline{H_\phi}$, such that $\dim \overline{H_\phi} = 9$. In particular, $9/|G(H_\phi^*)|$, which is a contradiction.

Suppose that $\dim A = 6$. Then, with the above notation, $\dim \overline{H_\phi} = 6$. Hence $\overline{H_\phi}^*$ contains a group-like Hopf subalgebra of dimension 3 or else a simple subcoalgebra of dimension 4, which again contradicts the fact that H_ϕ is of type $(1, 4; 4, 2)$ as an algebra.

Finally, if $\dim A = 18$, then $|G(A)| = 2$ and by [**M3**] A is commutative. It follows from [**AN2**, Corollary 3.9], $\dim V \leq [H_\phi : A] = 2$, for all irreducible H_ϕ-module V. This is impossible since H_ϕ is of type $(1, 4; 4, 2)$ as an algebra. This finishes the proof of the theorem. $\square$

CHAPTER 9

Dimension 40

9.1. Reduction of the problem

Let H be a nontrivial semisimple Hopf algebra of dimension 40 over k.

LEMMA 9.1.1. *The order of $G(H^*)$ is either 4, 8 or 20 and as an algebra H is of one of the following types:*
- $(1, 4; 2, 9)$,
- $(1, 4; 2, 5; 4, 1)$,
- $(1, 4; 2, 1; 4, 2)$,
- $(1, 8; 2, 8)$,
- $(1, 8; 2, 4; 4, 1)$,
- $(1, 8; 4, 2)$,
- $(1, 20; 2, 5)$.

We shall prove that the type $(1, 4; 2, 5; 4, 1)$ is impossible. See Lemma 9.1.5.

PROOF. Let $n = |G(H^*)|$. It follows from 1.1 that $n \neq 1, 5, 10$. The possibility $n = 2$ is discarded using Theorem 2.2.1. In the cases $n = 4, 8, 20$ the only possibilities excluded are the types $(1, 4; 3, 4)$ and $(1, 4; 6, 1)$; however, in this cases H has an irreducible character ψ of degree 3 (respectively 6), and decomposing the product $\psi\psi^*$ as a sum of irreducible characters gives a contradiction. This finishes the proof of the lemma. □

REMARK 9.1.2. (i) If H is of type $(1, 20; 2, 5)$, then H is not simple, by Corollary 1.4.3.

(ii) Let H be as in Lemma 9.1.1. It follows from Proposition 2.1.3 that there is a Hopf subalgebra $A \subseteq H$, with $\dim A = 8$.

LEMMA 9.1.3. *We have $H = R\#A$ is a biproduct where A is a semisimple Hopf algebra of dimension 8 and R is a 5-dimensional Yetter-Drinfeld Hopf algebra over A.*

PROOF. There are a Hopf subalgebra $A \subseteq H$ and a quotient Hopf algebra $q : H \to B$, such that $\dim A = \dim B = 8$. In particular, $\dim H^{\text{co } B} = 5$, and therefore $G(H) \cap H^{\text{co } B} = 1$ by [**NZ**].

Consider the restriction $q|_A : A \to B$. It will be enough to show that q is an isomorphism, or equivalently, that $A \cap H^{\mathrm{co}\, B} = k1$. Suppose on the contrary that $A \cap H^{\mathrm{co}\, B} = A^{\mathrm{co}\, B}$ is of dimension 2 or 4; see Lemma 1.3.4. Then there must exist $1 \neq g \in G(A) \cap A^{\mathrm{co}\, B}$, since the irreducible left coideals of A are of dimension 1 or 2. Then $2/\dim H^{\mathrm{co}\, B} = 5$, which is absurd. This contradiction finishes the proof of the lemma. □

LEMMA 9.1.4. *Suppose that H is simple and $|G(H)| = 4$. Then $R = k1 \oplus U$, where U is an irreducible left coideal of H of dimension 4. In particular, H is not of type $(1, 4; 2, 9)$ as a coalgebra.*

PROOF. Since $|G(H)| = 4$, A is not cocommutative. Suppose on the contrary that $R = k1 \oplus V_1 \oplus V_2$, where V_i is an irreducible left coideal of dimension 2. By Lemma 4.3.3 (i), $\rho(V_i) \subseteq kG(H) \otimes V_i$, $i = 1, 2$. Thus $\rho(R) \subseteq kG(H) \otimes R$ and $R \# kG(H)$ is a normal Hopf subalgebra of H (of index 2). □

LEMMA 9.1.5. *There exists no semisimple Hopf algebra H of type $(1, 4; 2, 5; 4, 1)$ as a coalgebra.*

PROOF. Suppose on the contrary that H is of type $(1, 4; 2, 5; 4, 1)$. Let $A \subseteq H$ be a Hopf subalgebra of dimension 8. For all irreducible H^*-characters χ such that $\deg \chi = 2$, we have that $G[\chi] \neq 1$, because H does not contain irreducible coideals of dimension 3. Also, the tensor product $\chi \chi'$ is irreducible for some irreducible characters χ and χ' of degree 2, since otherwise, there would be a Hopf subalgebra of dimension $4 + 4.5 = 24$, which is impossible. Therefore there exist $\chi, \psi \in X_2$ such that $G[\chi]$ is not contained in $G[\psi]$; c.f. Lemma 2.4.1. Thus $G[\chi]$ and $G[\psi]$ are distinct subgroups of order 2 of $G(H)$.

Since $|G[\chi]| = |G[\psi]| = 2$, then the orbits $G(H)\chi$ and $G(H)\psi$ are of order 2. Moreover, since $G[\chi'] = G[\chi]$, for all $\chi' \in G(H)\chi$ (because $G[\chi]$ is normal in $G(H)$), then $G(H)\chi$ and $G(H)\psi$ are disjoint orbits. Let $\tau \notin A$ be an irreducible character of degree 2. Then we have

$$\tag{9.1.6} \tau \tau^* = \sum_{a \in G[\tau]} a + \lambda,$$

for some $\lambda \in X_2$. By (9.1.6), we have $\lambda^* = \lambda$ and $G[\tau] \subseteq G[\lambda]$. Therefore, either $\lambda \in G(H)\tau$ or $\lambda \in A$.

CLAIM 9.1.7. *We have $\lambda \in A$.*

PROOF. Suppose on the contrary that $\lambda \in G(H)\tau$; hence, every orbit has a self-dual element. Moreover, if $g \in G(H)$ is such that $g\lambda = \tau$, then $\tau^* = \lambda g$ and therefore $G[\tau] = G[\lambda] \subseteq G[\tau^*]$. Hence $\tau^* \in G(H)\tau$. This implies, since each orbit has two elements, that

$\tau^* = \tau$. As before, we have decompositions $\tau^2 = \tau\tau^* = \sum_{a \in G[\tau]} a + \lambda$, where $\lambda \in G(H)\tau$ and if $\tau' \in G(H)\tau$, say $b\tau = \tau'$, then $\tau'\tau = b\tau^2 = b + ba + b\lambda$, and $b\lambda \in G(H)\tau$. Then the set $G(H) \cup G(H)\tau$ spans a standard subalgebra of $R(H^*)$ which corresponds to a Hopf subalgebra of dimension 12. This contradicts [**NZ**]; so we may conclude that λ is not in the orbit $G(H)\tau$. Therefore, we must have $\lambda \in A$ as claimed. □

Since $\chi\psi$ is irreducible, it follows from Lemma 2.4.1 that $m(\lambda, \psi^*\psi) = 0$. This contradicts Claim 9.1.7. Then the lemma follows. □

REMARK 9.1.8. Keep the notation in Lemma 9.1.3.
(i) Suppose that $R = k1 \oplus U$, where U is an irreducible left coideal of dimension 4 and let $\chi \in H$ be the corresponding character. It follows from Corollary 4.2.2, that $|G[\chi^*]| \leq 4$.
(ii) If $|G(H^*)| = |G(H)| = 8$, then H is not simple by [**N2**, Lemma 2.2.5]. So we may assume that A is not a group algebra of an abelian group. In particular, after dualizing if necessary, we may assume that A is not cocommutative.

LEMMA 9.1.9. *Suppose that H is of type $(1, 8; 2, 4; 4, 1)$ as a coalgebra. Then H is not simple.*

PROOF. By Remark 9.1.8 (ii), we may assume that $G(H)$ is not abelian of order 8. Consider the action by right multiplication of $G(H)$ on the set X_2 of irreducible characters of degree 2. Then either this action is transitive or else $|G[\chi^*]| = 4$, for all $\chi \in X_2$.

The first possibility implies, in view of Remark 2.4.3 (iii) that H has a Hopf subalgebra of dimension 24, which is impossible. Consider the second possibility. Since $G(H)$ is not abelian, $1 \neq [G; G] \subseteq G[\chi'] \cap G[\chi^*]$, for all $\chi', \chi \in X_2$. Therefore, by Theorem 2.4.2, also in this case there is a Hopf subalgebra of dimension 24, which is impossible. This proves the lemma. □

LEMMA 9.1.10. *Suppose that H is of type $(1, 8; 4, 2)$ as a coalgebra. Then H is not simple.*

PROOF. Suppose that H is simple. We have $\dim R(H^*) = 10$ and $G(H)$ is not abelian, by Remark 9.1.8 (ii). Let $e_0, e_1, e_2, e_3, f_1, f_2$ be orthogonal primitive idempotents in $kG(H)$ such that $\dim kG(H)e_j = 1$ and $\dim kG(H)f_j = 2$. Then $\dim He_j = 5$ and $\dim Hf_j = 10$, and moreover $Hf_1 \simeq Hf_2$.

We may assume that $|G(H^*)| \neq 8$. Then, by Lemma 9.1.4, $R^* \simeq k1 \oplus U$, where U is an irreducible left coideal of H^*, and *a fortiori* an irreducible Yetter-Drinfeld submodule of H^*.

By Proposition 1.5.2, there is a bijective correspondence between the irreducible Yetter-Drinfeld submodules of H appearing in $R^* = (H^*)^{\mathrm{co}\, kG(H)^*}$ and primitive idempotents $e \in R(H^*)$ such that $ee_0 \neq 0$.

Therefore we have a decomposition $e_0 = \Lambda + E_0$, where $\Lambda \in H$ is the normalized integral and E_0 is a primitive idempotent of $R(H^*)$ such that $\dim HE_0 = 4$. This gives at least 7 orthogonal idempotents in $R(H^*)$, implying that $R(H^*) \simeq k^{(6)} \times M_2(k)$ as an algebra. In particular, since f_1 and f_2 are not central in $kG(H)$, they correspond necessarily to the 4-dimensional simple component of $R(H^*)$.

We find that there must exist some $1 \leq j \leq 3$ such that e_j is not primitive in $R(H^*)$. By the Kac-Zhu Theorem, this implies the existence of a primitive idempotent $\Lambda \neq E \in R(H^*)$ such that $\dim HE = 1$. Then $G(H^*) \cap Z(H^*) \neq 1$ and the lemma follows. □

LEMMA 9.1.11. *Suppose that H is of type $(1, 8; 2, 8)$ as a coalgebra. Then H is not simple.*

PROOF. We have $H = R \# kG$, where $G = G(H)$ and $R = k1 \oplus V_1 \oplus V_2$, where V_i are irreducible left coideals of H such that $\dim V_i = 2$. By Theorem 2.2.1 we must have $G[\chi] \neq 1$, for all irreducible characters χ of degree 2. Then V_i is a subcoalgebra of R, by Proposition 4.2.1, and V_1 is not isomorphic to V_2 as left coideals of H. The group G permutes the set $G(V_1) \cup G(V_2)$ and we may also assume that the group homomorphism $G \to \mathbb{S}_4$ is injective. In particular, $G \simeq D_4$ and acts transitively on $G(V_1) \cup G(V_2)$.

Let $G_i = \{g : g.V_i \subseteq V_i\}$. Then $|G_i| \geq 4$ and in particular, $Z(G) \subseteq G_i$, $i = 1, 2$. By Lemma 4.7.1, if $Z(G)$ acts transitively on $G(V_i)$, then $|G_{V_i}| \leq 2$. Moreover, the action of $Z(G)$ is either transitive or trivial on $G(V_1)$. Observe in addition that V_1 generates R as an algebra. Hence, the last possibility implies that $Z(G) \subseteq Z(H)$ and H is not simple.

In the other case, we may conclude that $\rho(R) \subseteq kG_{V_i} \otimes R$ and since $|G_{V_i}| = 2$, $B = R \# kG_{V_i}$ is a Hopf subalgebra of dimension 10. By Remark 9.1.8 (ii), we may assume that $|G(H^*)| = 4$. Then $(H^*)^{\mathrm{co}\, B^*} = k1 \oplus kt \oplus V$, where $1 \neq t \in G(H)$ and V is an irreducible left coideal of dimension 2. Consider the possible decompositions of $(H^*)^{\mathrm{co}\, B^*}$ as a Yetter-Drinfeld submodule of H^*, as in Lemma 1.4.1. Since 3 does not divide $\dim H$, the dimensions of the irreducible summands are either 1 or 2. Decomposing each irreducible Yetter-Drinfeld summand as a sum of irreducible left coideals, implies that kt is necessarily a Yetter-Drinfeld submodule of H. Hence $t \in G(H) \cap Z(H)$ and H is not simple. This concludes the proof of the lemma. □

9.2. Main result

In view of the lemmas in the previous section, we may assume that H is of type $(1, 4; 2, 1; 4, 2)$ as an algebra and as a coalgebra. Moreover we have $H = R\#A$, where $A \simeq H_8$ and $R = k1 \oplus U$, for some irreducible left coideal U of dimension 4. We keep this notation along this section.

LEMMA 9.2.1. *If R is cocommutative then H is not simple.*

PROOF. We have $U \subseteq R$ is a subcoalgebra in ${}_A^A\mathcal{YD}$ and $\mathcal{S}_R U = U$. On the other hand, $Ut \simeq U$ for some $1 \neq t \in G(H) = G(A)$. Let $G(U) = \{x_1, \ldots, x_4\}$. By [**N3**, Lemmas 2.2.1 and 2.2.3] Ux_i is a left coideal of R and $Ux_i \simeq U(t.x_i)$; so if $t.x_i \neq x_i$, by counting dimensions, $x_i \in Ux_i$ implying that $1 \in U$, which is a contradiction. Therefore t acts trivially on R. The lemma will follow from the following claim:

CLAIM 9.2.2. *Let $A \to \operatorname{End} V$ be an action of $A \simeq H_8$ on a vector space V such that there exists $1 \neq t \in G(A)$ with $t|_V = \operatorname{id}_V$. Then $t'|_V = \operatorname{id}_V$ for all $t' \in G(A) \cap Z(A)$.*

PROOF. By [**M4**] we know that A is generated as an algebra by $G(A) = \{1, x, y, xy\}$ and an invertible element z such that $zx = yz$. Moreover, $G(A) \cap Z(A) = \{1, xy\}$. If $t = xy$, then there is nothing to prove. Thus, without loss of generality we may assume that $t = x$. In this case the relation $zx = yz$ plus the fact that z is invertible imply that also $y|_V = \operatorname{id}_V$. Then we have $xy|_V = \operatorname{id}_V$ as claimed. □

□

We now prove the main result of this chapter.

THEOREM 9.2.3. *Let H be a semisimple Hopf algebra of dimension 40. Then H is not simple.*

PROOF. We may assume that $H = R\#A$ is a biproduct, where $A \simeq H_8$ and $R = k1 \oplus U$. By Lemma 9.2.1 it will be enough to show that U is cocommutative. Suppose on the contrary that U is a simple coalgebra. Observe that $U\#A$ is a subcoalgebra of H; moreover, there exist $t, s \in G(H)$ such that $Ut \simeq U$ and Us is an irreducible coideal not isomorphic to U. In particular, $(U\#A) \cap C \neq 0$ and $(U\#A) \cap C' \neq 0$, where C and C' are the simple subcoalgebras of H containing U and Us, respectively. Thus, $U\#A = C \oplus C' \simeq M_4(k) \oplus M_4(k)$.

On the other hand, the assumption that U is simple implies that $U\#A \simeq U \otimes_\phi A$ for some invertible normalized 2-cocycle $\phi \in A \otimes A$; see [**Mo**, 7.3.1]. By [**M5**, Theorem 4.8 (1)] ${}_\phi A \simeq A$ as coalgebras, which gives $U\#A \simeq M_2(k)^{(4)} \oplus M_4(k)$. This is a contradiction, which implies that U is cocommutative and finishes the proof of the theorem. □

CHAPTER 10

Dimension 42

Let H be a semisimple Hopf algebra of dimension 42 over k. We shall assume that H is nontrivial. Suppose that H fits into an abelian extension

$$1 \to k^\Gamma \to H \to kF \to 1,$$

where Γ and F are finite groups. It follows from the results in [**N**, Section 4] that in this case H is isomorphic to one of the Hopf algebras $\mathcal{A}_7(2,3)$ or $\mathcal{A}_7(3,2) = \mathcal{A}_7(2,3)^*$ constructed in [**AN**, (2.3.1)]; indeed $\mathcal{A}_7(2,3)$ (respectively, $\mathcal{A}_7(3,2)$) fits into an extension as above, where Γ is the non-abelian group of order 14 and F is the cyclic group of order 3, (respectively, Γ is the non-abelian group of order 21 and F is the cyclic group of order 2).

For these Hopf algebras, $G(\mathcal{A}_7(3,2)) \simeq G(\mathcal{A}_7(2,3)) \simeq \mathbb{Z}_6$. As coalgebras, $\mathcal{A}_7(3,2)$ is of type $(1,6;2,9)$ and $\mathcal{A}_7(2,3)$ is of type $(1,6;3,4)$.

10.1. Possible (co)-algebra structures

Let H be a nontrivial semisimple Hopf algebra of dimension 42.

LEMMA 10.1.1. *The order of $G(H^*)$ is either 2, 6 or 14. As an algebra H is of one of the following types:*

- $(1,2;2,1;6,1)$,
- $(1,2;2,1;3,4)$,
- $(1,2;2,10)$,
- $(1,6;6,1)$,
- $(1,6;2,9)$,
- $(1,6;3,4)$,
- $(1,14;2,7)$.

PROOF. Let $n = |G(H^*)|$. A counting argument, using 1.1, shows that $n \neq 7, 21$, and the only possibilities are the prescribed ones if $n = 6$ or 14. If $n = 1$ we find that H must have an irreducible module of degree 2, which contradicts Corollary 2.2.3. If $n = 3$, then H is necessarily of type $(1,3;2,3;3,3)$, which is discarded by Theorem 2.2.1.

Suppose finally that $n = 2$. Then, either H is of one of the prescribed algebra types, or else H is of one of the following types:
$$(1,2;2,6;4,1), \qquad (1,2;2,2;4,2).$$
However, by Remark 2.2.2 (i), in these cases we have $G[\chi] = G(H^*)$, for all irreducible character χ of degree 2. By Theorem 2.4.2, H has a Hopf algebra quotient $H \to \overline{H}$, where $\dim \overline{H} = 2 + 6.4$ or $\dim \overline{H} = 2 + 2.4$, respectively. By [**NZ**], this is not possible. Then the lemma follows. $\square$

REMARK 10.1.2. (i) Since 12 does not divide the dimension of H, by Theorem 2.2.1, every irreducible character of degree 2 is stable under left multiplication by some group-like element $g \in G(H^*)$ of order 2.

(ii) It follows from Lemma 10.1.1 that $G[\chi]$ is nontrivial for all irreducible character χ of degree 6. Also, if $|G(H^*)| = 6$, then for all irreducible character χ of degree 3, $G[\chi]$ is of order 3.

(iii) It follows from Lemma 10.1.1 that $G(H^*)$ always contains a subgroup of order 2.

LEMMA 10.1.3. *(i) Suppose H is of type $(1,2;2,10)$ as a coalgebra. Then H is commutative.*

(ii) Suppose H is of type $(1,2;2,1;6,1)$ or $(1,2;2,1;3,4)$ as a coalgebra. Then H contains a Hopf subalgebra isomorphic to $k^{\mathbb{S}_3}$, where $\mathbb{S}_3$ is the symmetric group on 3 symbols.

PROOF. (i) This follows from Theorem 4.6.5, since by Remark 10.1.2 (iii), $|G(H^*)|$ is even.

(ii) In this case H has a unique simple subcoalgebra C of dimension 4, necessarily stable under left and right multiplication by $G(H)$. Therefore $A = kG(H) \oplus C$ is a Hopf subalgebra of H of dimension 6, which is not cocommutative; hence, in view of the classification of semisimple Hopf algebras of dimension 6, $A \simeq k^{\mathbb{S}_3}$. $\square$

LEMMA 10.1.4. *Suppose that H is of type $(1,14;2,7)$ as a coalgebra. Then H is commutative.*

PROOF. Note that the group $G(H^*)$ must be abelian by Proposition 1.2.6. Suppose that $G(H^*)$ is of order 2 or 6. By Lemmas 10.1.1 and 10.1.3, either there is a quotient Hopf algebra $p : H \to kG$, where G is a group of order 6, or else $G(H^*) \simeq \mathbb{S}_3$ is nonabelian of order 6. In the first case, a dimension argument implies that $\overline{H}^{cop} = k\Gamma$, where $\Gamma \subseteq G(H)$ is the only subgroup of order 7. Therefore, H fits into the (abelian) extension $1 \to k\Gamma \to H \to kG \to 1$, and the lemma follows from [**N**, Section 4]. In the second case, there is a projection $q : H \to kT \simeq k^T$, where $T \subseteq G(H^*)$ is the only subgroup of order 3; hence $\dim H^{co\,q} = 14$, and $\Gamma \subseteq H^{co\,q}$. A dimension argument shows that $G(H) \subseteq H^{co\,q}$ and thus $kG(H) = H^{co\,q}$. Therefore H fits into an abelian exact sequence $1 \to kG(H) \to H \to kT \to 1$, and the lemma follows also from [**N**, Section 4].

We claim that H^* is not of type $(1, 14; 2, 7)$ as a coalgebra. Indeed, in this case, we have $G(H) \simeq G(H^*)$ is abelian and $H \simeq R\#kG(H)$ is a biproduct, where R is of dimension 3. In particular, R is commutative and cocommutative, and it follows easily that the only subgroup F of order 7 of H acts trivially on R. Then F is central in H, and there is a central extension $0 \to kF \to H \to K \to 1$. Now, since every 6-dimensional semisimple Hopf algebra is trivial and since $|G(H^*)|$ is not divisible by 3, we find that K is necessarily isomorphic to a group algebra. In particular, this extension is abelian. This is a contradiction since $|G(H)| \neq 6$. This completes the proof of the lemma. □

LEMMA 10.1.5. *H is not of type $(1, 6; 6, 1)$ as a coalgebra.*

PROOF. Suppose on the contrary that H is of type $(1, 6; 6, 1)$ as a coalgebra. It follows from Lemmas 10.1.1, 10.1.3 and 10.1.4 that there is a quotient Hopf algebra $H \to A$, where A is of dimension 6. Hence, $A \simeq kG(H)$ and H is a biproduct $H \simeq R\#kG$, where $R \simeq H/H(kG)^+$ is a braided Hopf algebra over G of dimension 7, and $G = G(H)$.

As a left coideal of H, R decomposes as a direct sum $R = k1 \oplus V$, where V is an irreducible left coideal of dimension 6. In particular, V is a subcoalgebra of R in the category of Yetter-Drinfeld modules over G, and the smash coproduct coalgebra $V\#kG$ coincides with the unique simple subcoalgebra of H of dimension 36. Since $H^2(G, k^\times) = 1$, it follows from [**N2**, 1.3.1] that V is cocommutative and the action of G permutes the 6 distinct group-like elements. Thus V has a basis x_i, $1 \leq i \leq 6$, consisting of group-like elements of R.

By [**N3**, 2.2], Vx_i is a left coideal of R containing 1, and $Vx_i \simeq Vx_j$, for all $1 \leq i, j \leq 6$ such that $gx_i = x_j$, for some $g \in G$. Note that x_i does not belong to Vx_i: indeed, since x_i is invertible (with inverse $x_i^{-1} = \mathcal{S}_R(x_i)$), if $x_i = \sum_j x_j x_i$, then $1 \in \sum_j x_j \in V$, which is absurd.

Decomposing Vx_i into a direct sum of irreducible left coideals of R, we get $Vx_i = k1 \oplus \bigoplus_{l \neq i} kx_l$, where $g_i \neq g$, for all i. If $g \in G$ is such that $gx_i = x_j \neq x_i$, we have $gx_i \in Vx_i$. But then $x_j \in Vx_j \simeq Vx_i$. This is a contradiction. Then the lemma follows. □

LEMMA 10.1.6. *Suppose H is of coalgebra type $(1, 2; 2, 1; 6, 1)$ or $(1, 2; 2, 1; 3, 4)$. Then H is commutative.*

PROOF. We shall show that H fits into an abelian extension, and hence is trivial by [**N**, Section 4].

By Lemma 10.1.3, H has a Hopf subalgebra A isomorphic to $k^{\mathbb{S}_3}$. We claim that there is no Hopf algebra surjection $p : H \to k\mathbb{S}_3$. If this were the case, then $A \cap H^{cop} = k1$. Indeed, as a left coideal of H, H^{cop} decomposes as a direct sum or irreducible left coideals. Note that $G(A) = G(H)$ intersects trivially H^{cop} by [**NZ**]. If $A \cap H^{cop} \neq k1$, then there is a 2-dimensional irreducible left coideal U of A such that $U \subseteq H^{cop}$. Then $H^{cop} = k1 \oplus U \oplus W$, where W is a left coideal of H of dimension 4, such that W contains no one-dimensional left coideal of H. This is not possible and therefore $A \cap H^{cop} = k1$ as claimed. But then the restriction $p : k^{\mathbb{S}_3} \to k\mathbb{S}_3$ is an isomorphism, which is not possible. This establishes the claim. In particular, $|G(H^*)| \neq 2$.

We may then assume that $|G(H^*)| = 6$. Hence, by Lemma 10.1.5, H^* is of type $(1,6;3,4)$ or $(1,6;2,9)$ as a coalgebra. Dualizing the inclusion $A \subseteq H$, we get a Hopf algebra quotient $p : H^* \to k\mathbb{S}_3$. By [**NZ**], $G(H^*) \cap H^{cop} = k1$; so that the restriction $p : kG(H^*) \to k\mathbb{S}_3$ is an isomorphism. In particular, $G(H^*)$ is non-abelian and H^* is isomorphic to a biproduct $H^* \simeq R\#kG(H^*)$, where $\dim R = 7$. Let $\Gamma \subseteq G(H^*)$ be the unique subgroup of order 3.

Case I. H^* is of type $(1,6;3,4)$.

In this case, $R = k1 \oplus W_1 \oplus W_2$, where W_i are irreducible left coideals of H^* of dimension 3. Since $\Gamma = G[\chi]$ for all irreducible χ of degree 3, then $gW_i \simeq W_i \simeq W_ig$, for all $g \in \Gamma$, $i = 1, 2$.

CLAIM 10.1.7. *$\rho(W_i)$ is not contained in $k\Gamma \otimes W_i$.*

PROOF. Suppose on the contrary that $\rho(W_1) \subseteq k\Gamma \otimes W_1$. Let $\widetilde{R} := k[W_1]$; then $\widetilde{R}$ is a subalgebra and subcoalgebra of R which is stable under the action of Γ and $\rho(\widetilde{R}) \subseteq k\Gamma \otimes \widetilde{R}$. Therefore $\widetilde{A} = \widetilde{R}\#k\Gamma$ is a Hopf subalgebra of H^*, and $\dim \widetilde{A} = 3 \dim \widetilde{R} > 12$. Then $\dim \widetilde{A} = 21$, thus $\widetilde{A}$ is commutative and normal, and moreover the quotient $H^*/H^*\widetilde{A}^+$ is cocommutative, since $k\Gamma \subseteq \widetilde{A}$. Hence, $G(H^*)$ is abelian, which is a contradiction. This proves the claim. □

Fix $i = 1, 2$. We may assume that there exists $0 \neq w \in W_i$, such that $\rho(w) = a \otimes w$, where $a \in G(H^*)$ is an element of order 2. Let $1 \neq g \in \Gamma$, so that $0 \neq g.w \in W_i$ is homogeneous of degree $gag^{-1} \neq a$. Hence $W_i^{co\,G(H^*)} = 0$, and $R^{co\,G(H^*)} = k1$.

Dualizing, $H = R^*\#A$ and the subalgebra $(R^*)^A$ of invariants in R^* is one-dimensional, since $(R^*)^A \simeq \mathrm{Hom}^{G(H^*)}(R,k)$. By [**AN2**, Section 8], $(R^*)^A$ is isomorphic to the *Hecke algebra* of the pair A, H; that is, $(R^*)^A \simeq e_A H e_A$, where $e_A \in A$ is the normalized integral. By [**AN2**, Theorem 4.4] there is a bijection between irreducible representations of

$(R^*)^A$ and irreducible representations of H appearing in $\text{Ind}_A^H \epsilon_A$ with positive multiplicity. This is a contradiction, hence the lemma follows.

Case II. H^* is of type $(1, 6; 2, 9)$.

We shall use the action defined in (1.2.5).

CLAIM 10.1.8. *The action $\Gamma \times \Gamma \times X_2 \to X_2$, given by $(g, h).\chi = g\chi h^{-1}$ is freely transitive.*

PROOF. Let $\chi \in X_2$. It will be enough to show that the stabilizer $(\Gamma \times \Gamma)_\chi$ is trivial. Let $g, h \in \Gamma$ such that $g\chi h^{-1} = \chi$. Then $G[\chi] = G[g\chi h^{-1}] = gG[\chi]g^{-1}$. This implies that $g = 1$, because $G[\chi]$ is of order 2 and $G(H^*)$ is not abelian. Therefore $\chi h^{-1} = \chi$, and $h = 1$. This proves the claim. □

Observe that there exists $\chi \in X_2$ such that $\chi^* = \chi$, because $|X_2|$ is odd. Hence $\chi^2 = 1+a+\chi'$, where $(\chi')^* = \chi'$, and $G[\chi] = \{1, a\} = G[\chi']$. Note also that $\chi \neq \chi'$, since H has no quotient of dimension 6 which is not commutative. Write $\chi' = g\chi h^{-1}$, $g, h \in \Gamma$. Hence $G[\chi] = G[\chi'] = gG[\chi]g^{-1}$, implying $g = 1$. So $\chi' = \chi h^{-1}$ and $h \neq 1$. Thus $\chi h^{-1} = \chi' = (\chi')^* = h\chi^* = h\chi$; which implies that $h\chi h = \chi$. This contradicts Claim 10.1.8. The proof of the lemma is now complete. □

10.2. Classification

We shall now prove Theorem 3. In view of the previous results, we may assume that H is of type $(1, 6; 2, 9)$ or $(1, 6; 3, 4)$ as a coalgebra. The proof of the theorem of will follow from the next two lemmas.

LEMMA 10.2.1. *Suppose that H is of type $(1, 6; 2, 9)$ as a coalgebra. Then H is isomorphic to $\mathcal{A}_7(3, 2)$.*

PROOF. By the above, we also have $|G(H^*)| = 6$. Therefore the groups $G(H)$ and $G(H^*)$ are both abelian. This implies that $G(H)$ contains a unique subgroup F of order 2. In particular, $H \simeq R\#kF$ is a biproduct over F. The subgroup F necessarily stabilizes all 4-dimensional simple subcoalgebras. Therefore the quotient coalgebra $H/H(kF)^+ \simeq R$ is cocommutative. It follows from Proposition 4.6.1 that H fits into an abelian extension. This implies the lemma. □

LEMMA 10.2.2. *Suppose that H is of type $(1, 6; 3, 4)$ as a coalgebra. Then H is isomorphic to $\mathcal{A}_7(2, 3)$.*

PROOF. Also here we have $H \simeq R\#kG$ is a biproduct, where G is the unique subgroup of order 3 of $G(H)$. By Remark 10.1.2, G stabilizes all simple subcoalgebras of H of dimension 9. Then R is a cocommutative coalgebra and the lemma follows from Proposition 4.6.1. □

CHAPTER 11

Dimension 48

11.1. First reduction

Let H be a nontrivial semisimple Hopf algebra of dimension 48.

LEMMA 11.1.1. *The order of $G(H^*)$ is either* 2, 3, 4, 6, 8, 12, 16 *or* 24. *As an algebra, H is of one of the following types:*

- $(1, 2; 2, 3; 3, 2; 4, 1)$,
- $(1, 3; 3, 1; 6, 1)$,
- $(1, 3; 2, 9; 3, 1)$,
- $(1, 3; 3, 5)$,
- $(1, 4; 2, 2; 3, 4)$,
- $(1, 4; 2, 3; 4, 2)$,
- $(1, 4; 2, 7; 4, 1)$,
- $(1, 4; 2, 11)$,
- $(1, 4; 2, 2; 6, 1)$,
- $(1, 6; 2, 6; 3, 2)$,
- $(1, 8; 2, 2; 4, 2)$,
- $(1, 8; 2, 10)$,
- $(1, 8; 2, 6; 4, 1)$,
- $(1, 12; 2, 9)$,
- $(1, 12; 3, 4)$,
- $(1, 12; 6, 1)$,
- $(1, 16; 2, 8)$,
- $(1, 16; 4, 2)$,
- $(1, 24; 2, 6)$.

PROOF. The possibility $|G(H^*)| = 1$ is discarded by 1.1 and Theorem 2.2.1; see Corollary 2.2.3.

The only possibilities with $|G(H^*)| = 2$ are the types $(1, 2; 2, 7; 3, 2)$ and $(1, 2; 2, 3; 3, 2; 4, 1)$. In the first case, H cannot have Hopf algebra quotients of type $(1, 3; 3, 1)$. Therefore, by Remark 2.2.2 (i) every irreducible character of degree 2 is stable under left multiplication by $G(H^*)$; hence by Theorem 2.4.2 there is a quotient Hopf algebra of dimension 30, which is impossible.

Suppose that $|G(H^*)| = 16$. The only possibility excluded in the list is the type $(1, 16; 2, 4; 4, 1)$. Suppose on the contrary that H is of this type. Then H has four irreducible characters of degree 2 and one irreducible character of degree 4. In particular, the action of $G(H^*)$ on X_2 is transitive and $|G[\chi]| = 4$, for all irreducible character of degree 2. By Remark 2.4.3 (iii), there is a quotient Hopf algebra of dimension 32, which is impossible. The rest of the lemma follows from 1.1. □

REMARK 11.1.2. (i) If H is of type $(1, 24; 2, 6)$, then H is not simple, by Corollary 1.4.3.

(ii) Suppose that H is of type $(1, 4; 2, 3; 4, 2)$ as a coalgebra. By Theorem 2.4.2 the irreducible characters of degrees 1 and 2 give rise to a Hopf subalgebra of dimension 16.

(iii) If H is of type $(1, 8; 2, 2; 4, 2)$ as a coalgebra then the irreducible characters of degrees 1 and 2 give a Hopf subalgebra of dimension 16; see Remark 2.4.3 (iii).

(iv) Suppose that H is of type $(1, 2; 2, 3; 3, 2; 4, 1)$ as a coalgebra. Then H is not simple.

PROOF. We have that H does not contain Hopf subalgebras of dimension 12 [**F**]. On the other hand, there must exist an irreducible character χ of degree 2, such that $G[\chi] = 1$ (since otherwise there would be a Hopf subalgebra of dimension 14). By Theorem 2.2.1, H has a Hopf subalgebra of dimension 24 and then H is not simple. □

(v) Suppose that H is of type $(1, 4; 2, 2; 3, 4)$ as a coalgebra. Then H contains unique Hopf subalgebras $A_1 \subseteq A$ of dimension 6 and 12, respectively.

PROOF. The irreducible characters of degrees 1 and 2 give rise to a (unique) Hopf subalgebra of dimension 12.

Let $\psi_1, \ldots, \psi_4$ be the irreducible characters of degree 3, and let χ_1, χ_2 be the irreducible characters of degree 2. We have $\psi_1^* \psi_1 = 1 + \chi_1 + \psi + \psi'$, where ψ, ψ' are irreducible of degree 3; this implies that $\chi_i^* = \chi_i$, $i = 1, 2$. Thus $\psi_1 \chi_1 = \psi_1 + \psi_1 a$, where $1 \neq a \in G[\chi_1]$. In particular, $|G[\chi_1]| = |G[\chi_2]| = 2$.

Let ψ_i be any irreducible character of degree 3; thus $\psi_i = g\psi_1$, for some $g \in G(H)$. Hence $\psi_i^* \psi_i = \psi_1^* \psi_1 = 1 + \chi_1 + \psi + \psi'$.

Then $\chi_1^2 = 1 + a + \chi$, where χ is irreducible of degree 2. Comparing decompositions of $\psi_1 \chi_1^2$, we find that $\psi_1 \chi = \psi_1 + \psi_1 a$, implying that $m(\chi, \psi_1^* \psi_1) = 1$ and in turn that $\chi = \chi_1$. Then $G[\chi_1] \cup \{\chi_1\}$ give rise to a Hopf subalgebra A_1 of dimension 6. We have also $m(\chi_1, \chi_2 \chi_1) = m(\chi_2, \chi_1^2) = 0$. Therefore $m(\chi_2, \chi_2 \chi_1) = m(\chi_2, \chi_1 \chi_2) = m(\chi_1, \chi_2^2) > 0$, implying that $m(\chi_2, \chi_2 \chi_1) = m(\chi_1, \chi_2^2) = 1$. Thus $\chi_2^2 = b\chi_1$, $b \neq a \in$

$G(H)$ and $\chi_2^2 = 1+a+\chi_1$. In particular, A_1 is the only six-dimensional Hopf subalgebra of H as claimed. $\square$

Note that the set $\{\psi, \psi'\}$ is stable under the adjoint action of $G(H)$ and also under $*$. Also, if $\psi = \psi'$, then H is not simple; indeed, in this case, $G[\chi_1]$, χ_1, ψ and $a\psi = \psi a$ span a standard subalgebra corresponding to a Hopf subalgebra of dimension 24.

(vi) Suppose that H is of type $(1, 4; 2, 2; 6, 1)$ as a coalgebra. Then H contains a unique Hopf subalgebra A of dimension 12, which coincides with the sum of simple subcoalgebras of dimension 1 and 4.

LEMMA 11.1.3. *Suppose that H is of type $(1, 4; 2, 7; 4, 1)$ as a coalgebra. Then H is not simple.*

PROOF. By Proposition 2.1.3, H contains a Hopf subalgebra A of dimension 8. In particular, the group $G(H) = G(A)$ is not cyclic.

Let $\zeta \in H$ be the unique irreducible character of degree 4. Hence we have $g\zeta = \zeta = \zeta g$, for all $g \in G(H)$. Let $\lambda \in A$ be the unique irreducible character of degree 2. We claim that

$$\lambda \zeta = 2\zeta = \zeta \lambda.$$

Indeed, otherwise, there would exist $\chi \in X_2$ such that $m(\chi, \lambda \zeta) > 0$; hence $\lambda \zeta = \chi + \ldots$. Multiplying on the left by λ, and using that $\lambda^2 = \sum_{g \in G(H)} g$, we find that $\lambda \chi = \zeta$ is irreducible. This contradicts Lemma 2.4.1. Thus $\lambda \zeta = 2\zeta$, and then $\zeta \lambda = (\lambda \zeta)^* = 2\zeta^* = 2\zeta$. Hence the claim is established.

This implies that $AC = C = CA$. It follows from Corollary 3.2.5 and [**M5**, Theorem 4.8 (1)] that A is commutative. Suppose that $B \subseteq H$ is another Hopf subalgebra of dimension 8. Then the same argument applies, and thus $k[A, B]C = C = Ck[A, B]$. Hence $\dim k[A, B] = 16$ and the Hopf subalgebra $k[A, B]$ is normal in $k[C] = H$. Therefore H is not simple in this case.

Suppose next that $B \subseteq H$ is a Hopf subalgebra of dimension 6; we may assume that $k[A, B] = H$. We have $G(B) \cap Z(B) \neq 1$ and, since A is commutative, this group is central in $k[A, B] = H$. Hence H is not simple in this case.

Since H cannot contain Hopf subalgebras of dimension 32, there exist $\chi, \psi \in X_2$ such that $\chi^* \psi = \zeta$ is irreducible of degree 4. It follows from Lemma 2.4.1 that $G[\chi]$ and $G[\psi]$ are distinct subgroups of order 2 of $G(H)$. Write $\psi \psi^* = 1 + a + \tau$, where $\tau \in X_2$ and $G[\psi] = \{1, a\}$; thus $\tau = \tau^*$ and $G[\psi] \subseteq G[\tau]$. Similarly, $\chi \chi^* = 1 + b + \mu$, where $\mu \in X_2$ and $G[\chi] = \{1, b\}$; thus $\mu = \mu^*$ and $G[\chi] \subseteq G[\mu]$. In view of Lemma 2.4.1, we have $\mu \neq \tau$. Hence, after changing if necessary ψ and χ, we may assume $\tau \neq \lambda$.

Suppose $G[\tau] = G(H)$. Then τ and $G(H)$ span a standard subalgebra corresponding to a Hopf subalgebra of dimension 8, and we know H is not be simple in this case. Hence we may assume $|G[\tau]| = 2$. If $\tau = \psi g$, $g \in G(H)$, then $\tau^2 = \psi g(\psi g)^* = 1 + a + \tau$. Thus τ and $G[\tau]$ span a standard subalgebra corresponding to a Hopf subalgebra of dimension 6, and H is not simple.

Therefore we may assume that the orbits $\tau G(H)$ and $\psi G(H)$ are disjoint. Then the orbits of the right action of $G(H)$ on X_2 are

$$\lambda G(H), \quad \chi G(H), \quad \psi G(H), \quad \tau G(H),$$

and the only orbits with (left) stabilizer $G[\psi]$ are $\psi G(H)$ and $\tau G(H)$.

Note that if $\pi \in X_2$ is such that $m(\pi, \psi\zeta) > 0$, then $m(\zeta, \pi^*\psi) = m(\pi, \psi\zeta) > 0$. Therefore $G[\pi] \cap G[\psi] = 1$ and $\pi \in \chi G(H)$. Thus $\psi\zeta = \chi + \chi t + \zeta$, for some $t \in G(H)$. In particular, $m(\zeta, \zeta\psi^*) = m(\zeta, \psi\zeta) = 1$.

On the other hand, $\zeta\psi^* = \chi^*\psi\psi^* = \chi^* + \chi^* a + \chi^*\tau$. Hence $\chi^*\tau = \zeta$ is irreducible. Letting $X = \{\rho \in X_2 : \chi^*\rho = \zeta\}$, we have $X = \psi G(H) \cup \tau G(H)$. Moreover, we have $G(H)X = X = XG(H)$: the left hand side equality, because $G[g\rho] = G[\rho]$, for all $g \in G(H)$, and the only elements in X_2 with stabilizer $G[\psi]$ are in $\psi G(H) \cup \tau G(H) = X$.

Let $\sigma, \rho \in X' = X \cup \{\lambda\}$. Then the product $\sigma^*\rho \neq \zeta$. Note also that, for all $\rho \in X$, we have $m(\chi, \rho\zeta) = m(\chi g, \rho\zeta) = m(\zeta, \chi^*\rho) = 1$. Hence we have $\rho\zeta = \chi + \chi t + \zeta$, and thus $m(\rho, \zeta^2) = 1$, for all $\rho \in X$.

This allows us to write $\zeta^2 = \sum_{g \in G(H)} g + 2\lambda + \sum_{\rho \in X} \rho$. In particular, it follows that X' is closed under $*$. If $\sigma \in X$, then

$$\begin{aligned}
\sigma\zeta^2 &= \sum_{g \in G(H)} \sigma g + 2\sigma\lambda + \sum_{\rho \in X} \sigma\rho \\
&= (\sigma\zeta)\zeta = (\chi + \chi t + \zeta)\zeta = 2\chi\zeta + \zeta^2 \\
&= 2\sum_{\nu \in X} \nu + \sum_{g \in G(H)} g + 2\lambda + \sum_{\rho \in X} \rho.
\end{aligned}$$

This implies that for all $\sigma, \rho \in X$, the product $\rho\sigma$ decomposes as a sum of elements of $X' \cup G(H)$. Also, since $\lambda\zeta^2 = 2\zeta^2 = \zeta^2\lambda$, then all products $\lambda\sigma, \sigma\lambda$ decompose as sums of elements of X, for all $\sigma \in X$.

Hence $X' \cup G(H)$ spans a standard subalgebra of $R(H^*)$ which corresponds to a Hopf subalgebra of dimension 24. This proves the lemma. □

LEMMA 11.1.4. *Assume that H is of type $(1, 6; 2, 6; 3, 2)$ as a coalgebra. Then H is not simple.*

PROOF. Since the dimension of H is not divisible by 30, there must exist $\chi \in X_2$ such that $G[\chi] = 1$. By Theorem 2.2.1, H has a Hopf subalgebra A of type $(1, 3; 3, 1)$ as a coalgebra.

Let $\psi \in \widehat{H^*}$ be the unique irreducible character of degree 3 in A. Then we have $\psi^2 = 1 + a + a^2 + 2\psi$, where $G[\psi] = \{1, a, a^2\}$. Let $b \in G(H)$ of order 2, so that $b\psi = \psi' = \psi b$ is the remaining degree 3 irreducible character. We have $\psi'\psi$, $\psi\psi'$ belong to the span of $G(H) \cup \{\psi, \psi'\}$. Thus $G(H) \cup \{\psi, \psi'\}$ spans a standard subalgebra of H which corresponds to a Hopf subalgebra of dimension 24. This implies that H is not simple, as claimed. $\square$

LEMMA 11.1.5. *Assume that H is of type $(1, 3; 3, 5)$ as a coalgebra. Then H is not simple.*

PROOF. Let $\psi \in H$ be an irreducible character of degree 3. Decomposing the product $\psi\psi^*$, we see that $G[\psi] = G(H)$ is necessarily of order 3.

By Remark 3.2.7, the quotient coalgebra $H/H(kG(H))^+$ is cocommutative. In particular, we may assume that $|G(H^*)|$ is not divisible by 3: otherwise H would be a biproduct $H = R \# kG(H)$, with $R \simeq H/H(kG(H))^+$ cocommutative, implying that H is not simple, by Proposition 4.6.1.

By Lemma 11.1.1 and the previous lemmas, we may assume that there is a Hopf algebra quotient $H \to B$, where $\dim B = 4$; hence $\dim H^{\operatorname{co} B} = 12$, and by [**NZ**], $kG(H) \subseteq H^{\operatorname{co} B}$.

Let $H^{\operatorname{co} B} = kG(H) \oplus V_1 \oplus V_2 \oplus V_3$ be a decomposition of $H^{\operatorname{co} B}$ as a sum of irreducible left coideals of H. If V_1, V_2 and V_3 are pairwise isomorphic, then $H^{\operatorname{co} B}$ is a subcoalgebra of H and the map $H \to B$ is normal. Hence we may assume that V_1 appears with multiplicity 1 in $H^{\operatorname{co} B}$.

Suppose that V_i appears with multiplicity 1 in $H^{\operatorname{co} B}$. Then necessarily $gV_i = V_i = V_ig$, for all $g \in G(H)$, $i = 1, 2, 3$. Let $C_i \subseteq H$ be the simple subcoalgebra containing V_i. By Corollary 3.5.2 $kG(H)$ is normal in $k[C_i]$. Hence, we may assume that $\dim k[C_i] = 12$.

We claim that V_i appears with multiplicity 1, for all i. As above, implies that $G(H)$ is normal in $k[C_i]$ for all i. Since $k[C_1, C_2, C_3] = H$, it follows that $G(H)$ is normal in H and we are done.

To prove the claim we argue as follows. By Lemma 1.7.1, $(H^{\operatorname{co} B})^* \simeq \operatorname{Ind}_{B^*}^{H^*} 1$ as left H^*-modules. Hence, $\operatorname{Ind}_{B^*}^{H^*} 1 = \sum_{g \in G(H)} g + \psi_1 + 2\psi_2$, where $\psi_1 \neq \psi_2 \in H$ are irreducible characters of degree 3. In particular, $\psi_2^* = \psi_2$.

We may also assume that the Hopf subalgebra $k[C]$, where C is the subcoalgebra corresponding to ψ_2, is all of H; otherwise, $\dim k[C] = 12$

and $G(H)$ would be normal in $k[C]$, hence $G(H)$ would be normal in $H = k[C, C_1]$ (by dimension).

By Frobenius reciprocity, $\psi_2|_{B^*} = 2.1 + t$, where $1 \neq t \in G(B)$. Since $\psi_2^* = \psi_2$, then $t^2 = 1$. Let $A = k\langle t \rangle \subseteq B$. Consider the composition
$$\pi : H \to B \to \overline{B} := B/BA^+.$$
We have $\psi_2|_{\overline{B}} = \pi(\psi_2) = 3.1$. Applying again the Frobenius reciprocity, this gives $m(\psi_2, \mathrm{Ind}_{\overline{B}^*}^{H^*} 1) = 3 = \deg \psi_2$.

Therefore, by Lemma 1.7.1, $H^{\mathrm{co}\,\overline{B}}$ contains the simple subcoalgebra C corresponding to ψ_2. This is absurd, since it implies that $k[C] = H \subseteq H^{\mathrm{co}\,\overline{B}}$. This finishes the proof of the lemma. $\square$

LEMMA 11.1.6. *Assume that H is of type $(1, 3; 2, 9; 3, 1)$ as a coalgebra. Then H is not simple.*

PROOF. Let ψ be the unique irreducible character of degree 3. For all irreducible character μ of degree 2, we have $\mu\mu^* = 1 + \psi$. By Theorem 2.2.1, $G(H)$ and ψ span a standard subalgebra of $R(H)$, which corresponds to a commutative Hopf subalgebra of coalgebra type $(1, 3; 3, 1)$. In particular, the dimension of an irreducible H-module is at most $[H : A] = 4$. Thus H^* cannot be of type $(1, 3; 3, 1; 6, 1)$. Moreover, by previous results in this section, we may assume that either H^* is of type $(1, 3; 2, 9; 3, 1)$ as a coalgebra, or else there is a Hopf algebra quotient $H \to B$, where $\dim B = 4$.

In the last case, we have $\dim H^{\mathrm{co}\,B} = 12$ and $G(H) \subseteq H^{\mathrm{co}\,B}$, by [**NZ**]. Also, unless $A = H^{\mathrm{co}\,B}$ is normal in H, we may assume that $H^{\mathrm{co}\,B}$ contains an irreducible left coideal V of dimension 2. Therefore, $H^{\mathrm{co}\,B} = kG(H) \oplus \bigoplus_{g \in G(H)} gV \oplus W$ as a left coideal of H, where W is an irreducible left coideal of dimension 3. This implies that $G(H)\chi = \chi G(H)$, where χ is the character of V. Moreover, by Lemma 1.7.1, there exists $g_0 \in G(H)$ such that $\chi^* = g_0\chi$.

The relation $\chi^*\chi = 1 + \psi$ implies that $\psi\chi = \sum_{g \in G(H)} g\chi$. Also, for all $g, h \in G(H)$, we have $\chi h = h_0\chi$, for some $h_0 \in G(H)$ and therefore,
$$(g\chi)(h\chi) = gh_0\chi^2 = gh_0(g_0)^{-1}\chi^*\chi,$$
so that $(g\chi)(h\chi)$ belongs to the span of $G(H)$ and ψ. It follows that $G(H), \psi$ and $G(H)\chi$ span a standard subalgebra, corresponding to a Hopf subalgebra of dimension 24. Hence H is not simple in this case.

Suppose finally that H^* is of type $(1, 3; 2, 9; 3, 1)$ as a coalgebra. Then there is a projection $q : H \to B$, where $B = kG(B)$ is of dimension 12. The coalgebra structure of B^* implies that $G(B)$ has a unique normal subgroup of order 4. We have necessarily $H^{\mathrm{co}\,B} = k1 \oplus V$, where

V is an irreducible left coideal of dimension 3; whence $\dim q(A) = 3$ and $q(\psi) = \sum_{x \in G(q(A))} x$.

Since $|X_2|$ is odd, there exists an irreducible character χ of degree 2, such that $\chi^* = \chi$. Then $q(\chi) = a + b$, where $a, b \in G(B)$. The relation $\chi^2 = 1 + \psi$ implies that $G(q(A)) \subseteq \langle a, b \rangle$. We cannot have $a^2 = b^2 = 1$, since otherwise $\langle a, b \rangle$ would be contained in the unique subgroup of order 4 of $G(B)$. Hence $b = a^{-1}$, and $G(q(A)) = \langle a \rangle$ of order 3.

Let C be the simple subcoalgebra containing χ. Then $q(k[C]) = \langle a \rangle$; in particular, $k[C] \neq H$. In addition, $A \subseteq k[C]$, so that $12/\dim k[C]$. Hence, $\dim k[C] = 24$ and thus H is not simple. This finishes the proof of the lemma. $\square$

LEMMA 11.1.7. *Let H be of type $(1, 8; 2, 2; 4, 2)$ or $(1, 8; 2, 10)$ as a coalgebra. Assume H is simple. Then H contains a Hopf subalgebra of dimension* 16.

PROOF. If H is of type $(1, 8; 2, 2; 4, 2)$, the claim follows from Remark 11.1.2 (iii). So assume H is of type $(1, 8; 2, 10)$. For all $\lambda \in X_2$ we have $|G[\lambda]| = 2$ or 4.

Suppose $\chi \in X_2$ is such that $G[\chi] = G[\chi^*]$ of order 4. Let $A = kG[\chi]$ and C the simple subcoalgebra containing χ. We have $AC = C = CA$, whence A is normal in $k[C]$ by Proposition 3.2.6. Since $|G[\chi]| = 4$, A is normal in $kG(H)$ and then it is also normal in $K = k[C, G(H)]$. But $\dim K > 8$ is divisible by 8, and we are assuming that H is simple. Hence $\dim K = 16$ and we are done.

Assume first that $|G[\lambda]| = 4$ for all $\lambda \in X_2$. We claim that there exists $\chi \in X_2$ such that $G[\chi] = G[\chi^*]$, implying the lemma. To prove the claim, consider the action of $G(H) \times G(H)$ on X_2 given by $(g, h).\chi = g\chi h^{-1}$. The orbit of an element χ is $G(H)\chi G(H)$, so it has order 2 or 4, and clearly $G(H)\chi G(H)$ and $G(H)\chi^* G(H)$ are of the same order, for all χ. Also, because $G[\chi]$ is normal in $G(H)$, $G[\lambda] = G[\chi]$ for all $\lambda \in G(H)\chi G(H)$. Suppose on the contrary that $G[\chi] \neq G[\chi^*]$ for all $\chi \in X_2$. Then $G(H)\chi G(H)$ and $G(H)\chi^* G(H)$ are disjoint. Let $\alpha \in X_2$ such that α is not conjugate to χ nor to χ^* (such an α exists because $|X_2| = 10$). Then the orbits $G(H)\chi G(H)$, $G(H)\chi^* G(H)$, $G(H)\alpha G(H)$ and $G(H)\alpha^* G(H)$ are pairwise disjoint.

Then $|X_2| = 10 \geq 2|G(H)\chi G(H)| + 2|G(H)\alpha G(H)| \geq 8$, and there must exist β not in any of these orbits. Hence, by dimension, $G(H)\beta G(H) = G(H)\beta^* G(H)$. Thus $G[\beta] = G[\beta^*]$ and the lemma holds in this case.

Suppose next that $|G[\lambda]| = 2$ for some $\lambda \in X_2$. Then $\lambda\lambda^* = \sum_{g \in G[\lambda]} g + \chi$ for some $\chi \in X_2$ such that $\chi^* = \chi$. If $|G[\chi]| = 4$, then we are done. Otherwise, we may assume $\chi G(H)$ is the only orbit with 4 elements: if not, since $|X_2| = 10$, there would be a unique orbit $\alpha G(H)$ with stabilizer of order 4, and thus $G[\alpha^*] = G[\alpha]$ implying the lemma.

In particular, for all $g \in G(H)$, $g\chi g^{-1} = \chi a$, $a \in G(H)$, and therefore $G[g\chi g^{-1}] = gG[\chi]g^{-1} = G[\chi]$. That is, $G[\chi] \trianglelefteq G(H)$.

Also $\lambda = \chi t$, $t \in G(H)$, and $\chi^2 = \sum_{g \in G[\chi]} g + \chi$. Hence $G[\chi]$ and χ span a standard subalgebra corresponding to a Hopf subalgebra A of dimension 6. Moreover, $G[\chi]$ is central in A. Then $kG[\chi]$ is normal in $k[A, G(H)]$. But this implies $\dim k[A, G(H)] = 24$ or 48, and H is not simple in this case. $\square$

LEMMA 11.1.8. *Suppose that H is of type $(1, 8; 2, 6; 4, 1)$ as a coalgebra. Then H is not simple.*

PROOF. Let $\zeta \in C$ be the unique irreducible character of degree 4 in H. Let $X = \{\chi \in X_2 : m(\chi, \zeta^2) > 0\}$, so that $\zeta^2 = \sum_{g \in G(H)} g + \sum_{\chi \in X} n_\chi \chi + n\zeta$, where $n_\chi = m(\chi, \zeta^2) = m(\zeta, \chi\zeta) = 1$ or 2, $\chi \in X$. We have X is stable under $*$ and left and right multiplication by $G(H)$.

If $X = \emptyset$ then $G(H)$ and ζ span a standard subalgebra corresponding to a Hopf subalgebra of dimension 24. Hence H is not simple in this case. Thus we may assume $X \neq \emptyset$.

Let $\chi \in X$. We have $\chi\zeta = n_\chi \zeta + \sum \lambda$, where λ runs over the set $\Lambda = \{\lambda \in X_2 : m(\lambda, \chi\zeta) = m(\zeta, \lambda^*\chi) > 0\}$. Then $\deg \chi\zeta = 4n_\chi + 2|\Lambda|$. Note that $\Lambda G(H) = \Lambda$, so that $|\Lambda| = 0$ (iff $n_\chi = 2$) or 2 (iff $n_\chi = 1$).

If $|\Lambda| = 0$, we have $n_\chi = 2$. Since $g\chi$ appears in ζ^2 with the same multiplicity as χ, for all $g \in G(H)$. Then $|G[\chi]| = 4$ and $\zeta^2 = \sum_{g \in G(H)} g + 2\chi + 2t\chi$, for some $t \in G(H)$. In particular $G(H)\chi = \chi G(H)$. Also $\chi\zeta^2 = 2\zeta^2$. Then $G(H)$, $G(H)\chi$ and ζ span a standard subalgebra corresponding to a Hopf subalgebra of dimension 32. This is impossible by [**NZ**]. Hence $|\Lambda| = 2$; say $\Lambda = \{\lambda, \lambda a\}$, $a \in G(H)$. In particular, $|G[\lambda^*]| = |G[\lambda]| = 4$. Then $n_\chi = 1$, for all $\chi \in X$. So that $\zeta^2 = \sum_{g \in G(H)} g + \chi_1 + \chi_2 + \zeta$, with $X = \{\chi_1, \chi_2\}$, if $|X| = 2$, or $\zeta^2 = \sum_{g \in G(H)} g + \chi_1 + \chi_2 + \chi_3 + \chi_4$, with $X = \{\chi_1, \chi_2, \chi_3, \chi_4\}$, if $|X| = 4$.

Then, for each $\chi \in X$, $\chi\zeta = \zeta + \lambda + \lambda a$, and thus $\chi\zeta^2 = \zeta^2 + \lambda(1 + a)\zeta = \zeta^2 + 2\lambda\zeta$. Now since $\lambda^*\chi = \zeta$, then $\lambda\zeta = \lambda\lambda^*\chi \in G(H)\chi$. Hence $\chi\zeta^2$ belongs to the span of $G(H)$, ζ and X. Since this happens for all $\chi \in X$, then the irreducible summands of ζ^2 span a standard subalgebra corresponding to a Hopf subalgebra of dimension 32 or 24. By [**NZ**] the first case is impossible, and the second one implies that H is not simple. $\square$

LEMMA 11.1.9. *Suppose H is of type $(1,3;3,1;6,1)$ as a coalgebra. Then H is not simple.*

PROOF. The irreducible characters of degrees 1 and 3 span a standard subalgebra of $R(H)$, which corresponds to a Hopf subalgebra A of dimension 12. By [**F**] A is commutative. In particular, $\dim V \leq [H:A] = 4$, for all irreducible H-module V.

Suppose on the contrary that H is simple. Then there is no Hopf algebra quotient $H \to B$ with $\dim B = 16$: otherwise, by [**NZ**], $kG(H) = H^{\operatorname{co} B}$ is a normal Hopf subalgebra of H. In view of Lemma 11.1.1 and the previous results, it follows that $\dim V = 1, 2$ or 3 for all irreducible H-module V. It follows as well that $|G(H^*)| = 4$ or 12, and thus $G(H^*)$ contains a subgroup Γ of order 4.

Consider the Hopf algebra projection $H^* \to A^*$. Then $(H^*)^{\operatorname{co} A^*} \subseteq (H^*)^{\operatorname{co} q}$, where $q : H^* \to k^{G(H)}$ is the natural projection. By [**NZ**], we have $k\Gamma \subseteq (H^*)^{\operatorname{co} q}$. In particular, since $(H^*)^{\operatorname{co} q} \neq kG(H^*)$, $\Gamma = G(H^*) \cap (H^*)^{\operatorname{co} q}$ is a normal subgroup of $G(H^*)$, and it is therefore its only subgroup of order 4.

CLAIM 11.1.10. $(H^*)^{\operatorname{co} q}$ *contains no irreducible left coideal of dimension 3.*

PROOF. Suppose otherwise that $U \subseteq (H^*)^{\operatorname{co} q}$ is an irreducible left coideal of dimension 3. Hence H^* is of type $(1, 4; 2, 2; 3, 4)$ or $(1, 12; 3, 4)$ as a coalgebra, and there is a Hopf subalgebra $B \subseteq H^*$ with $\dim B = 12$, such that B is cocommutative or of type $(1, 4; 2, 2)$.

Since gU is not isomorphic to U, for all $g \in \Gamma$, the sum $\sum_{g \in \Gamma} gU$ is direct and by dimension, $(H^*)^{\operatorname{co} q} = k\Gamma \oplus \oplus_{g \in \Gamma} gU$. Hence, if H is simple, $(H^*)^{\operatorname{co} A^*} = k1 \oplus gU$, for some $g \in \Gamma$. This implies that $B^{\operatorname{co} A^*} = B \cap (H^*)^{\operatorname{co} A^*} = k1$, and therefore that $B \simeq A^*$ is cocommutative and $H^* \simeq R \# B$ is a biproduct, where R is a braided Hopf algebra over B of dimension 4. Proposition 4.4.6 implies that H is not simple in this case. Hence the claim follows. □

In view of the claim, we may assume that $(H^*)^{\operatorname{co} A^*} = k1 \oplus kt \oplus V$, where $1 \neq t \in G(H^*)$, and V is an irreducible left coideal of dimension 2. In particular, $tV = V = Vt$. So that, letting C be the simple subcoalgebra of H^* containing V, we find that t is central in the Hopf subalgebra $k[C]$, by Corollary 3.5.2. In particular, $k[C] \neq H^*$ and by [**NZ**], $\dim k[C] = 12$ or 8. In the first case, the restriction $q| : k[C] \to k^{G(H)}$ is surjective; then by [**F**] $k[C]$ is trivial, whence necessarily commutative. But this implies that $\dim U \leq [H^* : k[C]] = 4$, for all irreducible H-comodules U, against the assumption on the coalgebra type of H. Hence $\dim k[C] = 8$ and $k[C] \subseteq (H^*)^{\operatorname{co} q}$.

We have therefore a decomposition $(H^*)^{\operatorname{co} q} = k[C] \oplus V_1 \oplus V_2 \oplus V_3 \oplus V_4$, where V_i is an irreducible left coideal of dimension 2 of H^*, for all $i = 1, \ldots, 4$.

Note that V_i appears in $(H^*)^{\operatorname{co} q}$ with multiplicity 1, for all i. Otherwise, say $C_1 \subseteq (H^*)^{\operatorname{co} q}$, where C_1 is the simple subcoalgebra containing V_1. Then $k[C, C_1] \subseteq (H^*)^{\operatorname{co} q}$; but the inclusion $k[C] \subseteq k[C, C_1]$ is strict, so $k[C, C_1] = (H^*)^{\operatorname{co} q}$ by dimension restrictions, implying that H is not simple.

Let $\tau_i \in X_2$ be the irreducible character corresponding to V_i, $i = 1, \ldots, 4$, and let λ be the character of V. By Frobenius reciprocity $m(1, q(\tau_i \tau_i^*)) = 2$, thus $G[\tau_i] \neq \Gamma$, because $q(g) = 1$, for all $g \in \Gamma$. Hence $G[\tau_i]$ is a subgroup of order 2, for all $i = 1, \ldots, 4$. Similarly, we see that $m(\tau_i, \lambda \tau_i) = m(\lambda, \tau_i \tau_i^*) = 0$

Without loss of generality we may write $\tau_2 = h\tau_1$ and $\tau_4 = g\tau_3$, for some $h, g \in \Gamma \backslash \{1\}$. By the above, $\lambda \tau_1 = \tau_3 + \tau_4$. Write $q(\tau_1) = 1 + x$, where $1 \neq x$ is of order 3. Since $q(\lambda) = 2.1$, we find that $q(\tau_i) = 1 + x$, for all $i = 1, \ldots, 4$. This is impossible, since by Lemma 1.7.1, we must have $\tau_1^* = \tau_i$ for some i, and since $q(\tau_1^*) = 1 + x^{-1}$. This contradiction finishes the proof of the lemma. $\square$

LEMMA 11.1.11. *Suppose that H is simple. Then $|G(H)| \neq 12$.*

This discards the possibility that the types $(1, 12; 2, 9)$, $(1, 12; 3, 4)$ and $(1, 12; 6, 1)$ in Lemma 11.1.1 correspond to a simple Hopf algebra.

PROOF. Suppose on the contrary that $|G(H)| = 12$.

CLAIM 11.1.12. $|G(H^*)| = 12$.

PROOF. There is a Hopf algebra surjection $H^* \to B$, where $B \simeq k\mathbb{Z}_3$. Therefore, H^* does not contain any Hopf subalgebra A of dimension 16. Indeed, if this were the case, necessarily $(H^*)^{\operatorname{co} B} = A$, by [**NZ**]; hence H would not be simple.

Suppose that $|G(H^*)| \neq 12$. By previous results, H^* must be of type $(1, 4; 2, 2; 3, 4)$, $(1, 4; 2, 11)$ or $(1, 4; 2, 2; 6, 1)$ as a coalgebra.

Consider first the case where H^* is of one of the types $(1, 4; 2, 2; 3, 4)$ or $(1, 4; 2, 2; 6, 1)$; so that H^* contains a Hopf subalgebra A of dimension 12, which is of type $(1, 4; 2, 2)$ as a coalgebra. Since $kG(A) \subseteq (H^*)^{\operatorname{co} B}$, and $\dim(H^*)^{\operatorname{co} B} = 16$, then $(H^*)^{\operatorname{co} B} \cap A = kG(H^*)$. Let $\pi : H^* \to k^{G(H)}$ be the natural Hopf algebra projection. Then we have $(H^*)^{\operatorname{co} \pi} \subseteq (H^*)^{\operatorname{co} B}$ and $\dim(H^*)^{\operatorname{co} \pi} = 4$. Hence, $(H^*)^{\operatorname{co} \pi} = k1 \oplus W$, for some irreducible left coideal W of dimension 3. (In particular, H^* is not of type $(1, 4; 2, 2; 6, 1)$.)

Thus $(H^*)^{\mathrm{co}\,\pi} \cap A = k1$ and the restriction $\pi|_A : A \to k^{G(H)}$ is an isomorphism. This implies that H is a biproduct $H = R\#kG(H)$. Then the claim follows in this case from Proposition 4.4.6.

Suppose next that H^* is of type $(1,4;2,11)$. By Proposition 2.1.3, there is a Hopf algebra quotient $H \to B'$ where B' is of algebra type $(1,4;2,1)$; so that $\dim H^{\mathrm{co}\,B'} = 6$. This implies that H is not of type $(1,12;6,1)$.

By [**NZ**], any subgroup of order 3 of $G(H)$ is contained in $H^{\mathrm{co}\,B'}$. Hence, $G(H)$ contains a unique subgroup F of order 3, and F is the unique subgroup of $G(H)$ contained in $H^{\mathrm{co}\,B'}$.

Suppose H is of type $(1,12;3,4)$. Then $H^{\mathrm{co}\,B'} = kF \oplus V$, where V is an irreducible left coideal of H of dimension 3. Then we have $gV = V = Vg$, for all $g \in F$. Let C be the simple subcoalgebra of H containing V. By Corollary 3.5.2, kF is normal in $k[C]$.

Besides, since $H^{\mathrm{co}\,B'}$ is normal in H, we have $gCg^{-1} = C$, for all $g \in G(H)$. Since $k[C]$ and $G(H)$ necessarily generate H as an algebra, it follows that $k[C]$ is normal in H. This discards this type as the coalgebra structure of H.

Finally, suppose H is of type $(1,12;2,9)$. Then we must have $H^{\mathrm{co}\,B'} = kF \oplus \oplus_{g \in F} gV$, where V is an irreducible left coideal of H of dimension 2. This is not possible, since $\dim H^{\mathrm{co}\,B'} = 6$. Thus H is not type $(1,12;2,9)$ and the proof is complete. $\square$

Claim 11.1.12 implies that H is not of type $(1,12;6,1)$ as a coalgebra: indeed, in this case, $H^{\mathrm{co}\,kG(H^*)} \subseteq kG(H)$ and therefore $kG(H^*)$ is a normal Hopf subalgebra of H^*.

Since $|G(H)| = |G(H^*)| = 12$, then $H = R\#k\mathbb{Z}_3$ is a biproduct. In particular, $G(H)$ contains a unique (normal) subgroup of order 4. There is in addition a Hopf algebra surjection $H \to B$, where $\dim B = 4$. Thus if $F \subseteq G(H)$ is a subgroup of order 3, then $F \subseteq H^{\mathrm{co}\,B}$. This shows that $G(H)$ also contains a unique (normal) subgroup of order 3. Thus $G(H)$ is abelian. This implies that H is not of type $(1,12;3,4)$, in view of Proposition 4.6.1, since in this case R will be a cocommutative coalgebra, by Remark 3.2.7.

It remains to discard the case where H is of type $(1,12;2,9)$. Let $\Gamma \subseteq G(H)$ be the unique subgroup of order 4. Then there is an irreducible character λ of degree 2, such that $\Gamma \cup \{\lambda\}$ spans a standard subalgebra of $R(H^*)$, corresponding to a Hopf subalgebra A of dimension 8.

By [**NZ**], we must have $A \subseteq R$. Also, since R is not a Hopf subalgebra, R can contain only one simple subcoalgebra of dimension 4, which implies that this subcoalgebra (and hence all of A) is stable under the action of $\mathbb{Z}_3$ by conjugation.

Hence $B = k[A, \mathbb{Z}_3]$ contains A as a normal Hopf subalgebra, and thus $B \neq H$. Since $\dim B$ is divisible by 3 and $8 = \dim A$, we get $\dim B = 24$. This implies that H is not simple. The lemma follows. $\square$

11.2. Further reductions

In this section we further reduce the possibilities for the (co)algebra structure of an eventual simple H.

LEMMA 11.2.1. *Suppose that H is of type $(1,4;2,11)$ as a coalgebra. Then we have*

(i) There exists an 8-dimensional non-cocommutative Hopf subalgebra $A_0 \subseteq H$.

(ii) Assume in addition that H is simple and contains a Hopf subalgebra A of dimension 12. Then neither A nor A_0 is commutative. In this case, there exists a twist $\phi \in kG(H) \otimes kG(H)$ such that $|G(H_\phi)| = 12$, 24 or 48. In particular, H_ϕ is not simple.

PROOF. Part (i) follows from Proposition 2.1.3. We show part (ii). Note that a Hopf subalgebra of H of dimension 12 is necessarily of coalgebra type $(1,4;2,2)$. Let $B = k[A_0, A]$ be the subalgebra generated by A_0 and A, so that B is a Hopf subalgebra of H. Since $8 = \dim A_0$ divides the dimension of B and also $3/\dim A/\dim B$, then 24 divides the dimension of B. H being simple by assumption, we may assume that $B = H$. In other words, A_0 and A generate H as an algebra.

Also $G(A_0) = G(A) = G(H)$ is not cyclic.

Suppose that A is commutative. We know that there exists a central group-like element $1 \neq g \in G(A_0)$. Since A is commutative, and $G(A_0) = G(A)$, then g commutes with A. Therefore g is central in $k[A_0, A] = H$, contradicting the simplicity of H. A similar argument shows that A_0 cannot be commutative. The last statement of the lemma follows from Proposition 5.2.1. $\square$

LEMMA 11.2.2. *Suppose $H = R\#A$ is a biproduct, where $\dim A = 16$. Then H is not simple.*

PROOF. We have $\dim R = 3$, thus R is commutative and cocommutative. As a left coideal of H, we may assume that $R = k1 \oplus V$, where V is an irreducible left coideal of dimension 2.

If A is cocommutative, then the lemma follows from Proposition 4.4.6. So we may assume A is not cocommutative. By Lemma 4.3.3, $\rho(R) \subseteq kG(A) \otimes R$.

11.2. FURTHER REDUCTIONS

It is not difficult to see that there must exist a normal Hopf subalgebra B of A such that $kG(A) \subseteq B$ and $\dim B = 8$. Thus $\rho(R) \subseteq B \otimes R$ and $R \# B$ is a normal Hopf subalgebra (of index 2) of H. □

LEMMA 11.2.3. *Suppose that H contains a Hopf subalgebra A with $\dim A = 16$. If there is a quotient Hopf algebra $q : H \to B$, with $\dim B = 16$, then H is not simple.*

PROOF. We have $\dim H^{\operatorname{co} B} = 3$. Hence, by Lemma 1.3.4, $A \cap H^{\operatorname{co} B} = k1$, thus $H = R \# A$ is a biproduct. The lemma follows from Lemma 11.2.2. □

LEMMA 11.2.4. *Suppose that H is of coalgebra type $(1, 4; 2, 2; 3, 4)$ or $(1, 4; 2, 2; 6, 1)$. Assume in addition that there is a Hopf algebra quotient $q : H \to B$, with $\dim B = 16$. Then H is not simple.*

PROOF. We consider first the case where H is of type $(1, 4; 2, 2; 6, 1)$. Let ψ be the unique irreducible character of degree 6, and let χ_1, χ_2 be the irreducible characters of degree 2. Then $\chi_1, \chi_2 \in A$, where A is the unique Hopf subalgebra of dimension 12 of H; see Remark 11.1.2 (vi). We shall show that A is normal in H.

It is not hard to see that

$$(11.2.5) \qquad \psi^2 = \sum_{g \in G(H)} g + 2\chi_1 + 2\chi_2 + 4\psi.$$

Necessarily, we must have $H^{\operatorname{co} B} = k1 \oplus V$, where V is an irreducible left coideal of H of dimension 2. In particular, $H^{\operatorname{co} B} = A^{\operatorname{co} B}$ and $q(A)$ is a four-dimensional Hopf subalgebra of B.

We first claim that B is of type $(1, 4; 2, 3)$ as a coalgebra. To prove this, we consider the possible decompositions of $q(\psi) = \psi|_{B^*}$ into a sum of irreducible characters in B. It follows from Frobenius reciprocity, together with equation (11.2.5), that $q(\psi) = \lambda_1 + \lambda_2 + \lambda_3$, where λ_i are pairwise distinct irreducible characters of degree 2 in B. Hence B must be of the prescribed coalgebra type.

Therefore, $q(A) = kG(B)$; since this is the unique Hopf subalgebra of dimension 4 in B.

On the other hand, it follows from the classification results in [**K**], that $kG(B)$ is normal in B. For the sake of completeness, we give a proof of this fact in what follows. We may assume that B is not commutative. By [**K**, 3.3], B fits into a cocentral abelian extension

$$1 \to K \to B \to kF \to 0,$$

where $F = \langle t : t^2 = 1 \rangle \simeq \mathbb{Z}_2$ and K is a commutative Hopf algebra of dimension 8. In particular, $kG(B)$ is contained in K and therefore $G(B) = G(K)$ is central K.

Let $e = \sum_{x \in G(B)} g$. As an algebra, H is a smash product $H = K\#kF$, with respect to an action $\rightharpoonup: F \times K \to K$ by Hopf algebra automorphisms. Hence, the action of $t \in F$ permutes the elements of $G(B)$, and therefore $t \rightharpoonup e = e$.

Note that H is generated as algebra by K and $T := 1\#t$. It follows from the above discussion that
$$Te = (t \rightharpoonup e)\#t = e\#t = Te.$$
Hence $e \in Z(H)$, which proves the desired fact.

Consider the sequence of surjective Hopf algebra maps
$$H \xrightarrow{q} B \xrightarrow{q'} B',$$
where $B' = B/B(kG(B))^+$. Since $q(A) = kG(B)$, we have $A \subseteq H^{\operatorname{co} q'q}$. Thus, by dimension, $A = H^{\operatorname{co} q'q}$ and it is a normal Hopf subalgebra, as claimed.

Now consider the case where H is of type $(1, 4; 2, 2; 3, 4)$. We shall keep the notation in Remark 11.1.2 (v).

For all irreducible character ψ_i of degree 3, we have relations

(11.2.6) $$\psi_i^* \psi_i = 1 + \chi_1 + \psi + \psi',$$

and $\chi_1 \psi_i = \psi_i + a\psi_i$, where ψ, ψ' are fixed irreducible of degree 3, $G[\chi_1] = \{1, a\}$, and $\chi_1 \in A_1$, where A_1 is the unique six-dimensional Hopf subalgebra of H.

Observe first that $\psi' = \psi g$, for some $g \in G(H)$. We shall show that $g \in G[\chi_1]$. So that the irreducible characters $1, a, \chi_1, \psi, a\psi$, span a standard subalgebra of $R(H^*)$ corresponding to a Hopf subalgebra of dimension 24, whence H is not simple in this case.

We have $H^{\operatorname{co} B} = k1 \oplus V$, where V is an irreducible left coideal of dimension 2, and necessarily $V \subseteq A_1$, since $\dim A_1$ does not divide $\dim B$. Hence, $q(\chi_1) = (\chi_1)|_{B^*} = 1 + \alpha$, where $1 \neq \alpha \in G(B)$. Since $a\chi_1 = \chi_1$, and $q(a) \neq 1$, then $q(a) = \alpha$.

On the other hand, $m(1, q(\psi_i)) = m(\psi_i, \operatorname{Ind}_{B^*}^{H^*} 1) = 0$, for all $i = 1, \ldots, 4$. The relation (11.2.6) implies that $m(1, q(\psi_i^* \psi_i)) = 2$.

This implies that $q(\psi) = \beta + \lambda$ and $q(\psi') = \beta' + \lambda'$, for some $\beta, \beta' \in G(B)$, and $\lambda, \lambda' \in X_2(B)$. By Frobenius reciprocity, β, β' and α are pairwise distinct elements of $G(B)$.

Using again equation (11.2.6) for $\psi_i = \psi$, we find that $\lambda^* \lambda = 1 + \beta + \beta' + \alpha$; hence $\{1, \beta, \beta', \alpha\} = G[\lambda^*]$ is a subgroup of $G(B)$. Therefore $\beta\alpha = \beta'$.

Using Frobenius reciprocity, we have
$$\psi a = (\operatorname{Ind}_{B^*}^{H^*} \beta) a = \operatorname{Ind}_{B^*}^{H^*}(\beta a|_{B^*}) = \operatorname{Ind}_{B^*}^{H^*}(\beta \alpha) = \operatorname{Ind}_{B^*}^{H^*}(\beta') = \psi'.$$

11.2. FURTHER REDUCTIONS

This establishes the claim and finishes the proof of the lemma. □

LEMMA 11.2.7. *Suppose that H contains a Hopf subalgebra A with $\dim A = 16$. Then we have:*

(i) If H is simple, then H is of type $(1, 4; 2, 11)$ as an algebra, and there exists a normalized 2-cocycle $\phi \in kG(H^*) \otimes kG(H^*)$ such that $(H^*)_\phi$ is not simple.

(ii) Assume in addition that A is cocommutative. Then H is not simple.

By part (ii), if H is of type $(1, 16; 2, 8)$ or $(1, 16; 4, 2)$, then H is not simple. Therefore, if H is simple and contains a Hopf subalgebra of dimension 16, H is of type $(1, 8; 2, 2; 4, 2)$, $(1, 8; 2, 10)$, $(1, 4; 2, 3; 4, 2)$ or $(1, 4; 2, 11)$ as a coalgebra.

PROOF. (i) By previous lemmas, $|G(H^*)| \neq 2, 3, 6, 12$ and there is no Hopf subalgebra $B \subseteq H^*$ with $\dim B = 16$.

Consider the projection $\pi : H^* \to A^*$; we may assume $(H^*)^{\operatorname{co} A^*} = k1 \oplus V$, where V is a left coideal of H^* of dimension 2. It follows from Lemma 11.1.1 that $G[\chi] \neq 1$, where $\chi = \chi_V$ is the character corresponding to V. Then $|G[\chi]| = 4$ or 2.

By Theorem 2.5.1, H^* contains a Hopf subalgebra B of dimension $3|G(H^*)|$. We may assume that $|G(H^*)| = 4$ or 8, and therefore, $\dim B = 12$ or 24.

Since H is simple, $\dim B = 12$, and therefore $|G(H^*)| = 4$. It follows from Lemma 11.2.3 and the results of the previous section, that H^* is of type $(1, 4; 2, 2; 3, 4)$, $(1, 4; 2, 2; 6, 1)$ or $(1, 4; 2, 11)$ as a coalgebra. In view of Lemma 11.2.4, H^* is of type $(1, 4; 2, 11)$ as a coalgebra.

Since H^* contains a Hopf subalgebra A of dimension 12, by Lemma 11.2.1, there exists a normalized 2-cocycle $\phi \in kG(H^*) \otimes kG(H^*)$ such that $|G(H^*_\phi)| = 12, 24$ or 48. Then H^*_ϕ is not simple. This proves (i).

(ii) We may assume that $|G(H)| = 16$ and H is simple. As in the proof of part (i), it follows that H^* contains a Hopf subalgebra B of dimension 12. Consider the dual projection $q : H \to B^*$; we have $\dim H^{\operatorname{co} B^*} = 4$. Also, by a dimension argument using [**NZ**], the kernel of the restriction of q to $G(H)$ must be of order 4. Thus $H^{\operatorname{co} B^*} \subseteq kG(H)$ and q is normal. Hence H is not simple in this case. □

We summarize the results of this section in the following corollary.

COROLLARY 11.2.8. *If H is simple, the possible coalgebra types for H and H^* must be among the following:*

$$(1, 4; 2, 2; 3, 4), \quad (1, 4; 2, 2; 6, 1), \quad (1, 4; 2, 11)$$
$$(1, 4; 2, 3; 4, 2), \quad (1, 8; 2, 2; 4, 2), \quad (1, 8; 2, 10).$$

Moreover, either H or H^* must be of one of the types listed in the first row, and if the type of H is in the second row then the type of H^* is $(1,4;2,11)$. □

11.3. Main result up to cocycle twists

We shall prove in this section that semisimple Hopf algebras of dimension 48 are not simple, *up to a cocycle twist*.

LEMMA 11.3.1. *Suppose H is simple. Then H admits no Hopf algebra quotient $H \to B$, where B is a cocommutative Hopf algebra of dimension 12.*

PROOF. We shall consider separately the coalgebra types listed in Corollary 11.2.8.

CLAIM 11.3.2. *Suppose $H \to B$ is a Hopf algebra quotient where B is cocommutative and $\dim B = 12$. Then H^* is of type $(1,4;2,2;3,4)$ or $(1,4;2,2;6,1)$ as a coalgebra.*

PROOF. By assumption $B^* \subseteq H^*$ is a commutative Hopf subalgebra of dimension 12. If H^* is not of the prescribed types, then there is a Hopf subalgebra $A \subseteq H^*$ of dimension 8. Since H is simple, we may assume $k[A, B^*] = H$. On the other hand, there exists $1 \neq g \in Z(A) \cap G(A)$. If $g \in B^*$, then g is central in $k[A, B^*] = H$ and H is not simple. If $g \notin B^*$, then $|G(H^*)| = 8$ and $G(B^*) \subseteq G(H^*)$ is a normal subgroup (of index 2). Then $kG(B^*)$ is normal in $k[G(H^*), B^*] = H$. □

Suppose first that H is of type $(1,4;2,2;6,1)$ as a coalgebra. The lemma follows in this case from 1.1, since a cocommutative quotient Hopf algebra B of dimension 12 of H is the same as a commutative Hopf subalgebra B^* of index 4 of H^*, and since H^* has an irreducible module of dimension $6 > [H^* : B^*]$.

Suppose that, as a coalgebra, H is of one of the types

$$(1,4;2,3;4,2), \quad (1,8;2,2;4,2), \quad (1,8;2,10).$$

Then H contains a Hopf subalgebra of dimension 16 and H^* is of type $(1,4;2,11)$ as a coalgebra. Then the lemma follows from Lemma 11.2.1.

Suppose next that H is of type $(1,4;2,2;3,4)$. Recall from Remark 11.1.2 (v), that $G(H)$ and the irreducible characters χ_1 and χ_2 of degree 2 give rise to a Hopf subalgebra A of dimension 12, and for all irreducible characters $\psi_1, \ldots, \psi_4$ of degree 3 we have $\psi_i^* \psi_i = 1 + \chi_1 + \psi + \psi'$, where ψ and ψ' are fixed elements in X_3. We also have $\chi_1^2 = 1 + a + \chi_1$, where $\{1, a\} = G[\chi_1] \subseteq G(H)$.

Suppose that H has a Hopf algebra quotient $q : H \to B = kG(B)$, where $\dim B = 12$. We may assume that $m(1, q(\psi_i)) = 0$, for all $i = 1, \ldots, 4$. Otherwise, we would have a decomposition $H^{\mathrm{co}\,B} = k1 \oplus W$, where W is an irreducible left coideal of dimension 3, implying that the restriction $q : A \to B$ is an isomorphism, and thus that A is cocommutative, which is absurd.

We claim that also $m(1, q(\chi_1)) = 0$. If not, since $q(\chi_1) = \chi_1|_{B^*}$, then $m(1, q(\chi_1)) = 1$ by Frobenius reciprocity. Then $q(\chi_1) = 1 + g$, where $1 \neq g \in G(B)$. The relation $\chi_1^2 = 1 + a + \chi_1$, implies that $q(a) = g$.

Let $1 \neq t \in G(H) \cap H^{\mathrm{co}\,B}$. Then $q(t) = 1$, and $(\mathrm{Ind}_{B^*}^{H^*} 1)t = \mathrm{Ind}_{B^*}^{H^*} 1$; so that $t = a \in G[\chi_1]$. This is a contradiction. Therefore $m(1, q(\chi_1)) = 0$, as claimed.

The relation $\psi_i^* \psi_i = 1 + \chi_1 + \psi + \psi'$ implies that $m(1, q(\psi_i^* \psi_i)) = 1$. Write $q(\psi_i) = \sum_{s \in G(B)} n_s s$. Then $\sum_s n_s = \deg \psi_i = 3$, and $n_s = m(s, q(\psi_i)) = m(\psi_i, \mathrm{Ind}_{B^*}^{H^*} s) \leq 1$. Hence, the set of all $s \in G(B)$ such that $n_s \neq 0$ has at least 3 elements. Now we have

$$q(\psi_i^* \psi_i) = q(\psi_i)^* q(\psi_i) = \sum_{s,u \in G(B)} n_s n_u s^{-1} u,$$

whence $m(1, q(\psi_i^* \psi_i)) \geq 3$. This is again a contradiction. Therefore the lemma is established in this case.

It remains to consider the case where H is of type $(1, 4; 2, 11)$. Suppose that there is Hopf algebra quotient $H \to B$, where $B = kG(B)$ is a cocommutative Hopf algebra of dimension 12. By Corollary 11.2.8, B^* is of type $(1, 4; 2, 2)$ as a coalgebra. Hence, $G(B)$ has a normal subgroup of order 3, and there is a normal Hopf subalgebra $B_0 \subseteq B$, with $\dim B_0 = 3$.

We may write $H^{\mathrm{co}\,B} = k1 \oplus kt \oplus V_1$, where V_1 is an irreducible left coideal of dimension 2, and $1 \neq t \in G(H)$. By Corollary 3.5.2, t is central in $k[C_1]$, where C_1 is the simple subcoalgebra containing V_1.

Consider the sequence of surjective Hopf algebra maps

$$H \xrightarrow{q} B \xrightarrow{q'} B',$$

where $B' = B/BB_0^+$. We have $\dim H^{\mathrm{co}\,B'} = 12$, and there exists an irreducible left coideal U of H, not isomorphic to V_1 such that $m(U, H^{\mathrm{co}\,B'}) > 0$.

Let $\chi \in H$ and $C \subseteq H$ be, respectively, the irreducible character and the simple subcoalgebra corresponding to U. Decompose $q(\chi) = \chi|_{B^*}$ in the form $q(\chi) = g + h$, where $g, h \in G(B)$. Then the restriction of q induces an epimorphism $q : k[C] \to k\langle g, h\rangle \subseteq kG(B)$.

Similarly, the restriction of $q'q$ induces an epimorphism $q : k[C] \to k\langle g', h'\rangle \subseteq kG(B')$, where $q'q(\chi) = g' + h'$. In particular, g' and h' are the natural projections of g and h in $G(B')$.

By Frobenius reciprocity, we may assume that $g' = 1$. Hence
$$q'q(k[C]) = k\langle h'\rangle \neq B',$$
because, by Claim 11.3.5, $G(B')$ is not cyclic. In particular $k[C] \neq H$.

On the other hand, since $g' = 1$, then $g \in B_0$ is of order 3 (because $m(1, q(\chi)) = 0$). Hence 3 divides the dimension of $k[C]$. Since H is simple, then $\dim k[C] = 12$ or 6. Also, $\dim k[C]^{\text{co } B} = 2$, and it turns out that $1 \neq t$ is a central group-like in $k[C]$.

Since this happens for every irreducible constituent U of $H^{\text{co } B'}$, not isomorphic to V, it follows that t is central in the Hopf subalgebra generated by all simple subcoalgebras intersecting $H^{\text{co } B'}$. This implies that t is central in H. This finishes the proof of the lemma. □

LEMMA 11.3.3. *Suppose that H is simple. Then there exists an invertible normalized 2-cocycle $\phi \in kG(\widetilde{H}) \otimes kG(\widetilde{H})$ such that $G(\widetilde{H}_\phi) \simeq G$, where G is the group defined in 5.2. Here, $\widetilde{H}$ is one of the Hopf algebras H or H^*.*

Recall from Chapter 5 that G is the semidirect product $G = F \rtimes \Gamma$, where $F = \langle a : a^3 = 1\rangle$, and $\Gamma = \langle s, t : s^2 = t^2 = sts^{-1}t^{-1} = 1\rangle$, corresponding to the action by group automorphisms of Γ on F defined on generators by $s.a = a^{-1}$ and $t.a = a^{-1}$.

Observe that, by Lemma 11.1.1, $\widetilde{H}_\phi$ is of one of the types
$$(1, 12; 2, 9), \quad (1, 12; 3, 4), \quad (1, 12; 6, 1).$$
In particular, $\widetilde{H}_\phi$ is not simple.

PROOF. We shall show that there exist a Hopf subalgebra $A \subseteq \widetilde{H}$ such that $A \simeq \mathcal{A}_0$. Thus there is an invertible normalized 2-cocycle $\phi \in kG(\widetilde{H}) \otimes kG(\widetilde{H})$ such that $A_\phi \simeq kG$, in view of Proposition 5.2.1. In particular, $|G(\widetilde{H}_\phi)|$ is divisible by 12, where $\widetilde{H}$ is either H or H^*. Thus $|G(\widetilde{H}_\phi)| = 12, 24$ or 48. The following claim implies that indeed $|G(\widetilde{H}_\phi)| = 12$. Hence, $G(\widetilde{H}_\phi) \simeq G$.

CLAIM 11.3.4. *Suppose $|G(\widetilde{H}_\phi)| = 24$ or 48. Then H is not simple.*

PROOF. If $|G(\widetilde{H}_\phi)| = 24$, then $kG(\widetilde{H}_\phi)$ is a normal Hopf subalgebra of $\widetilde{H}_\phi$. On the other hand we must have $G(\widetilde{H}) \subseteq G(\widetilde{H}_\phi)$. Then $\phi^{-1} \in kG(\widetilde{H}_\phi) \otimes kG(\widetilde{H}_\phi)$. Therefore, in view of Lemma 5.4.1,

$B = (kG(\widetilde{H}_\phi))_{\phi^{-1}}$ is a normal Hopf subalgebra of $\widetilde{H}$ (because it has index 2).

Assume now that $|G(\widetilde{H}_\phi)| = 48$; that is, $\widetilde{H}_\phi = k\Gamma$, where $\Gamma = G(\widetilde{H}_\phi)$ is a group of order 48. For such a group Γ, either $Z(\Gamma) \neq 1$ or Γ contains a normal subgroup of order 16, that necessarily contains $G(\widetilde{H})$. In any case, $\widetilde{H}$ (and thus H) is not simple, as claimed. □

Since H is simple by assumption, we know that H is of one of the coalgebra types listed in Corollary 11.2.8. Eventually taking $\widetilde{H} = H^*$, we may further assume that H has one of the following coalgebra types:

$$(1,4;2,2;3,4), \quad (1,4;2,2;6,1), \quad (1,4;2,11).$$

CLAIM 11.3.5. *The group $G(H)$ is not cyclic.*

PROOF. If H is of type $(1,4;2,11)$, the claim follows from Lemma 11.2.1 (i). If H is of type $(1,4;2,2;6,1)$, then H contains a simple subcoalgebra C of dimension 36 such that $gC = C = Cg$, for all $g \in G(H)$, and the quotient coalgebra $C/C(kG(H))^+$ is of dimension 9. If $G(H)$ is cyclic, then any twisted group algebra $k_\alpha G(H)$ is cocommutative, but this contradicts Corollary 3.2.5, because $C/C(kG(H))^+$ cannot have 4 isomorphism classes of simple comodules. Hence the claim follows also in this case.

It remains to consider the type $(1,4;2,2;3,4)$. Suppose on the contrary that $G(H)$ is cyclic. By Corollary 11.2.8, there is a quotient Hopf algebra $q: H \to Q$ of dimension 4, so that $\dim H^{\mathrm{co}\,q} = 12$. Keep the notation in Remark 11.1.2 (v). Considering the eventual decomposition of $A^{\mathrm{co}\,q}$, $A_1^{\mathrm{co}\,q}$, we distinguish the following two possibilities for $H^{\mathrm{co}\,q}$:

(a) $H^{\mathrm{co}\,q} = k1 \oplus V \oplus W_1 \oplus W_2 \oplus W_3$, where V is an irreducible left coideal of dimension 2, and W_i's are irreducible left coideals of dimension 3, $i = 1, 2, 3$.

(b) $H^{\mathrm{co}\,q} = k1 \oplus kt \oplus V \oplus V' \oplus W_1 \oplus W_2$, where $t \in G(H)$, V, V' are irreducible coideals of dimension 2, and W_i's are irreducible left coideals of dimension 3, $i = 1, 2$.

Note that in this case t is an element of order 2 in $G(H)$, and since this group is cyclic by assumption, necessarily $t = a \in G[\chi_1]$.

Case (a). In this case the restriction of q to $kG(H)$ induces an isomorphism $kG(H) \simeq Q$. Hence $G(H^*)$ contains a cyclic subgroup of order 4, and by Corollary 11.2.8, we may assume that H^* is also of type $(1,4;2,2;3,4)$ as a coalgebra. Hence there is a quotient Hopf algebra $H \to B$, where $\dim B = 12$ and $H^{\mathrm{co}\,B} \subseteq H^{\mathrm{co}\,q}$ is a left coideal of dimension 4. Thus we must have a decomposition $H^{\mathrm{co}\,B} = k1 \oplus W$, where W is an irreducible left coideal of dimension 3. Let $\psi_W \in H$

be the character of W. Then $g\psi_W g^{-1} = \psi_W$, for all $g \in G(H)$, and $(\psi_W)^* = \psi_W$ by Lemma 1.7.1. As in Remark 11.1.2 (v) we have a decomposition

$$(11.3.6) \qquad (\psi_W)^*\psi_W = 1 + \chi_1 + \psi + \psi'.$$

By Frobenius reciprocity, $m(1, (\psi_W)|_B) = 1$, hence $(\psi_W)|_B = 1 + \lambda$, where $\lambda \in B$ is a (not necessarily irreducible) character such that $m(1, \lambda) = 0$; then $m(1, (\psi_W)^*(\psi_W)|_B) \geq 2$. On the other hand,

$$m(1, (\psi_W^*\psi_W)|_B) = 1 + m(1, (\chi_1)|_B) + m(1, \psi|_B) + m(1, \psi'|_B)$$
$$= 1 + m(1, \psi|_B) + m(1, \psi'|_B),$$

and we may assume that $m(1, \psi) > 0$, whence $\psi_W = \psi$.

This implies that $\psi^* = \psi$. Then also $(\psi')^* = \psi'$, in view of the relation (11.3.6). Write $\psi' = g\psi$, $1 \neq g \in G(H)$. Therefore $g^2 = 1$, and since $G(H)$ is cyclic $g = a \in G[\chi_1]$. But this implies that the irreducible characters $G[\chi_1], \chi_1, \psi, \psi'$ span a standard subalgebra which corresponds to a Hopf subalgebra of H of dimension 24; hence H is not simple in this case.

Case (b). Let ψ_i be the character of W_i, $i = 1, 2$. Then $a\psi_1 = \psi_2 \neq \psi_1$, because aW_1 is not isomorhic to W_1 and it is contained in $H^{\text{co } q}$. Moreover, $\psi_1 \in \{\psi, \psi'\}$, since otherwise, we would have $\psi' = a\psi$, implying as before that H contains a Hopf subalgebra of dimension 24.

Then we may assume that $\psi_1 = \psi$. Write $\psi' = g\psi$, $1 \neq g \in G(H)$. If $\psi^* = \psi$, then also $(\psi')^* = \psi'$ and therefore $g^2 = 1$, whence $g = a$ since $G(H)$ is cyclic. If $\psi^* = \psi'$, then $\psi' = \psi_2$, by Lemma 1.7.1. Thus, in any case, $\psi' = a\psi$, implying as before, that H contains a Hopf subalgebra of dimension 24. This proves that if H is simple, then the group $G(H)$ is not cyclic, as claimed. $\square$

Suppose that H has a Hopf subalgebra A of dimension 12. By Lemma 11.3.1, H^* has no cocommutative quotient of dimension 12. Therefore, A is not commutative. Since $G(H)$ is not cyclic, then $A \simeq \mathcal{A}_0$ by [**F**]. By Proposition 5.2.1 (i), there exists a 2-cocycle $\phi \in kG(A)^{\otimes 2} = kG(H)^{\otimes 2}$ such that $\mathcal{A}_0 \simeq (kG)_\phi$. This establishes the lemma in this case.

We may therefore assume that H is of type $(1, 4; 2, 11)$ and contains no Hopf subalgebra of dimension 12, by Lemma 11.2.1. In particular, H^* contains no Hopf subalgebra of dimension 16, by Theorem 2.5.1. We may also assume that H^* is not of type $(1, 4; 2, 2; 3, 4)$ nor $(1, 4; 2, 2; 6, 1)$, since otherwise we are done by letting $\widetilde{H} = H^*$.

By Lemma 11.2.7 (i), we may suppose that H^* is also of type $(1, 4; 2, 11)$. Then there is a quotient $q : H \to B$, where $\dim B = 8$ and

B is not commutative. We may assume that $H^{\mathrm{co}\,B} = k1 \oplus kt \oplus V_1 \oplus V_2$, where $t \in G(H)$ and V_i is an irreducible coideal of dimension 2, such that V_i is not isomorphic to V_2. Let $\tau_i \in X_2$ and $C_i \subseteq H$ be the irreducible character and simple subcoalgebra corresponding, respectively, to V_i, $i = 1, 2$.

Then $m(\tau_i, \mathrm{Ind}_{B^*}^{H^*} 1) = 1$, and therefore $q(\tau_i) = \tau_i|_{B^*} = 1 + g_i$, where $1 \neq g_i \in G(B)$. Moreover, the subgroups $\langle g_1 \rangle$, $\langle g_2 \rangle$ are of order at most 4. By construction, q induces by restriction a surjective Hopf algebra map $q : k[C_i] \to k\langle g_i \rangle$; in particular, $k[C_i] \neq H$.

If $k[C_i]^{\mathrm{co}\,B} = k1 \oplus V_i$, respectively $k1 \oplus kt \oplus V_1 \oplus V_2$, then we have $\dim k[C_i, G(H)] = 12$, respectively $\dim k[C_i] = 12$. Hence we are done in these cases. If otherwise, $k[C_i]^{\mathrm{co}\,B} = k1 \oplus kt \oplus V_i$, for all $i = 1, 2$, Then $tV_i = V_i = V_i t$ and by Corollary 3.5.2, t is central in $k[C_i]$. Thus t is central in $k[C_1, C_2]$; and we may assume that $k[C_1, C_2] \neq H$. Since 6 divides $\dim k[C_1, C_2]$, then we may even assume that $\dim k[C_1, C_2] = 12$ and we are done in view of Lemma 11.2.1. □

11.4. Main result

In this section we shall prove our main result in dimension 48. For this we shall first study the normal Hopf subalgebras in a semisimple Hopf algebra K with $|G(K)| = 12$. In what follows, we shall denote by G the group considered in Lemma 11.3.3.

THEOREM 11.4.1. *Let H be a semisimple Hopf algebra of dimension 48. Then H is not simple.*

PROOF. By Lemma 11.3.3, we may assume that for $\widetilde{H} = H$ or H^*, there exists an invertible normalized 2-cocycle $\phi \in kG(\widetilde{H}) \otimes kG(\widetilde{H})$ such that $\widetilde{H}_\phi$ is of one of the types

$$(1, 12; 2, 9), \quad (1, 12; 3, 4), \quad (1, 12; 6, 1),$$

and $G(\widetilde{H}_\phi) \simeq G$, where G is as in 5.2.

By Lemma 11.1.11, $\widetilde{H}_\phi$ is not simple. Note that $kG(\widetilde{H}) = (kG(\widetilde{H}))_\phi$ is a cocommutative Hopf subalgebra of dimension 4 of $\widetilde{H}_\phi$. Note also that, because $\widetilde{H} = \widetilde{H}_\phi$ as algebras, we may assume that $|G(\widetilde{H}_\phi^*)|$ is divisible by 4. On the other hand, $\phi^{-1} \in kG(\widetilde{H}) \otimes kG(\widetilde{H})$ is a normalized 2-cocycle for $\widetilde{H}_\phi$.

Let K be a semisimple Hopf algebra of dimension 48 such that $G(K) \simeq G$. In view of Lemmas 11.4.4, 11.4.5 and 11.4.7 below, at least one of the following conditions hold:

(11.4.2) K has a nontrivial central group-like element;

(11.4.3) $G(K)$ is contained in a normal Hopf subalgebra of K.

Thus, if $J \in kG(K) \otimes kG(K)$ is an invertible 2-cocycle, the twisted Hopf algebra K_J is not simple, by Lemma 5.4.1 and Corollary 5.4.2. This implies the theorem, applied to $K = \widetilde{H}_\phi$ and $J = \phi^{-1}$. □

In the rest of this section K will be a semisimple Hopf algebra of dimension 48 such that $G(K) \simeq G$; that is, $G(K) \simeq G = F \rtimes \Gamma$, and thus G contains a unique abelian subgroup M of order 6, $M = F \times Z$, where $Z = Z(G)$ is of order 2.

Also, $A \subseteq H$ will be a proper normal Hopf subalgebra. Our aim is to show that at least one of the conditions (11.4.2) or (11.4.3) hold.

Hence, we may assume in what follows that $\dim A \neq 2$. Indeed, by Corollary 1.4.3, if $\dim A = 2$, then $A \subseteq kG(K) \cap Z(K)$. Thus, in this case, K verifies condition (11.4.2).

On the other hand, note that we cannot have $A \cap kG(K)$ of dimension 4: this would imply that $A \cap kG(K) = kL$, where L is a *normal* subgroup of order 4 in G (because $A \cap kG(K)$ is invariant under the adjoint action of G), which contradicts the structure of G.

Hence $\dim A = 4, 8, 16$ are impossible.

Therefore, the possibilities for $\dim A$ are 3, 6, 12 and 24. Moreover, if $\dim A = 12$ or 24, we may assume that K is of type $(1, 12; 3, 4)$. Otherwise, (that is, if K is of type $(1, 12; 2, 9)$ or $(1, 12; 6, 1)$), since $\dim A \cap kG(K) \neq 4$, necessarily $G(K) \subseteq A$. See Lemma 6.1.1.

LEMMA 11.4.4. *Assume K is of type $(1, 12; 2, 9)$ as a coalgebra. Then at least one of the conditions (11.4.2) or (11.4.3) hold.*

PROOF. By Proposition 2.1.3, K contains a Hopf subalgebra K_0 of dimension 8. We may assume that $k[K_0, F] = K$; if not, $\dim k[K_0, F] = 24$ and $k[K_0, F]$ is normal in K, whence we are done because $G(K) = G \subseteq k[K_0, F]$.

For all $g \in F$, gK_0g^{-1} is an 8-dimensional Hopf subalgebra of K. If $gK_0g^{-1} = K_0$ for some $1 \neq g \in F$, then $G(K_0)$ would be a normal subgroup of order 4 of G, which is a contradiction. Thus the conjugation action of F gives rise to 8-dimensional Hopf subalgebras K_0, K_1, K_2 and $G(K_i)$, $0 \leq i \leq 2$, are distinct subgroups of order 4 of G.

We may further assume that $F \subseteq k[K_0, K_1, K_2]$ and therefore that $k[K_0, K_1, K_2] = K$. Otherwise $G(k[K_0, K_1, K_2])$ would be of order 4 and since $k[K_0, K_1, K_2]$ is normal in $K = k[K_0, K_1, K_2, F]$, then this group would be normal in $G(K)$, which is not possible. Note that for each $i = 0, 1, 2$, $G(K) = \langle G(K_i), F \rangle$.

We may assume K contains a normal Hopf subalgebra A with $\dim A = 3$ or 6. If $\dim A = 3$, then $A = kF$, and in this case, since $kG(K_i)$ is normal in K_i, then $\text{ad}_{K_i} kG(K) = kG(K)$. Thus $kG(K)$ is normal in $K = k[K_0, K_1, K_2]$. If on the other hand, $\dim A = 6$, then because $\dim K/KA^+ = 8$, we have $kF \subseteq A$. Hence A is cocommutative and arguing as before we get that $kG(K)$ is normal in K. This finishes the proof of the lemma. □

LEMMA 11.4.5. *Assume K is of type $(1, 12; 3, 4)$ as a coalgebra. Then at least one of the conditions (11.4.2) or (11.4.3) hold.*

PROOF. In this case may assume that $\dim A = 3, 6, 12$ or 24. We may further assume that K contains a normal Hopf subalgebra of dimension 3; that is, $\dim A = 3$. To see this, we argue as follows. Suppose that $\dim A = 12$. Then A is of type $(1, 3; 3, 1)$ as a coalgebra, and A is commutative. Also, $G(A)$ is the unique (normal) subgroup of order 3 of $G(K) = G$. Thus $kG(A)$ is a 3-dimensional normal Hopf subalgebra in $k[A, G] = K$.

If $\dim A = 6$, then $A = kN$, where N is a normal subgroup of order 6 of $G(K)$. Consider the Hopf algebra quotient $K \to K/KA^+$, since $\dim K/KA^+ = 8$, there is a Hopf algebra quotient $K \to K/KA^+ \to B$, where $\dim B = 4$. We have $A \subseteq K^{\text{co }B}$, and unless $kG(K)$ is normal in K, $K^{\text{co }B} = A \oplus U \oplus V$, where U and V are irreducible left coideals of K of dimension 3; since $NK^{\text{co }B} = K^{\text{co }B}$, we get that U and V are not isomorphic.

In addition, $xV \simeq V \simeq Vx$ and $xU \simeq U \simeq Ux$, for all $x \in F$, because $G[\chi_U]$ and $G[\chi_V]$ are necessarily of order 3. Let C_U and C_V be the simple subcoalgebras of K containing U and V, respectively. In view of Corollary 3.5.2, kF is normal in $k[C_V, C_U]$. Therefore, since it is also normal in $kG(K)$, then kF is normal in $k[C_V, C_U, G(K)] = K$.

Finally, if $\dim A = 24$, we may assume that A is of type $(1, 6; 3, 2)$ as a coalgebra; see Lemma 6.1.1. In particular, $G(A)$ is a normal subgroup of G of order 6.

Let $A_0 \subseteq A$ be a normal Hopf subalgebra. If $\dim A_0 = 12$, then $kF \subseteq Z(A_0)$, and it follows that kF is normal in $K = k[A_0, G(K)]$. If, on the other hand, A_0 is cocommutative, then $\dim A_0 = 2, 3$ or 6.

In the first case, $A_0 = kZ$ is the only subgroup of order 2 contained in $G(A)$. Hence kZ is central in A and thus also in $k[A, G(K)] = K$. If $\dim A_0 = 3$, then $A_0 = kF$ and kF is normal in $k[A, G(K)] = K$. If $\dim A_0 = 6$, then $A_0 = kN$ is normal in $k[A, G(K)] = K$ and K has a normal Hopf subalgebra of dimension 3, be the above.

We may therefore assume that $\dim A = 3$, as claimed. Then $A = kF$, since F is the only subgroup of order 3 in $G(K)$. In particular, $F = G[\chi]$ for all irreducible character χ of degree 3. Hence, the quotient Hopf algebra K/KA^+ is cocommutative; see Remark 3.2.7. Say $K/KA^+ \simeq kN$, where N is a group of order 16. Since $kF \simeq k^F$, then K fits into the abelian extension

(11.4.6) $$1 \to k^F \to K \to kN \to 1.$$

We shall prove in what follows that in this case, $G(K) \cap Z(K) \neq 1$. Hence K satisfies condition (11.4.2).

Consider the matched pair (F, N) associated to the extension, with the actions $\triangleleft : F \times N \to F$ and $\triangleright : F \times N \to N$. By [**N**, Lemma 1.1.7], the exact sequence dual to (11.4.6) gives rise to an exact sequence of groups

$$1 \to \widehat{F} \to G(K) \to N^F,$$

where N^F is the subgroup of invariants in N under the action $\triangleright$. Since $|G(K)| = 12$, N^F contains a subgroup N_1 of order 4 of N.

On the other hand, the action $\triangleleft$ permutes the set $F \backslash \{1\}$; hence, there is a subgroup $N_0 \subseteq N$ such that N_0 acts trivially on F and $|N_0| = 8$. We have $N_0 \cap N_1 \neq 1$. Let $1 \neq x \in N_0 \cap N_1$.

As a Hopf algebra, $K \simeq k^{F^\tau} \#_\sigma kN$ is a bicrossed product, corresponding to the actions $kN \otimes k^F \to k^F$ obtained from $\triangleleft$, and $kF \otimes k^N \to k^N$ obtained from $\triangleright$, and certain 2-cocycles $\sigma : N \times N \to (k^F)^\times$ and $\tau : F \times F \to (k^N)^\times$. Moreover, by [**N**, Lemma 1.2.5], we may assume that $\tau = 1$.

The compatibility conditions between $\triangleleft$ and $\triangleright$ imply that F acts on N_0 by group automorphisms through $\triangleright$. Hence $A_0 = k^F \otimes kN_0$ is a Hopf subalgebra of K of dimension 24.

Since $x \in N_1$, then F acts trivially on x and $x \in G(K)$. Also, since $x \in N_0$, then x acts trivially in F, and thus x commutes with k^F [**M**].

Since F acts on N_0 by group automorphisms, this action preserves the center of N_0. If N_0 is not abelian, then $|Z(N_0)|$ is of order 2, and F acts trivially on $Z(N_0)$. Hence we may assume that $x \in Z(N_0)$. Therefore, $1 \neq x \in G(A_0) \cap Z(A_0)$ is an element of order 2. Clearly, the same conclusion holds if N_0 is abelian, since in this case any $x \in N_1 \cap N_0$ commutes with N_0.

On the other hand, we may assume that A_0 is of type $(1, 6; 3, 2)$ as a coalgebra, whence $G(A_0) \cap Z(A_0) \supseteq Z(G(K))$, because A_0 is normal in K. Therefore $Z(G(K))$ is central in $k[A_0, G(K)] = K$. This proves the claim. The proof of the lemma is now complete. □

LEMMA 11.4.7. *Assume K is of type $(1, 12; 6, 1)$ as a coalgebra. Then at least one of the conditions (11.4.2) or (11.4.3) hold.*

PROOF. In this case may assume that $\dim A = 3$ or 6. We claim that if K has a Hopf algebra quotient $K \to B$ with $\dim B = 16$, then $kG(K)$ is normal in K. Hence the lemma follows when $\dim A = 3$.

Observe that the quotient $q : K \to B$ is necessarily normal, since $kF = K^{\operatorname{co} B}$ by [**NZ**]. In order to establish the claim, we shall follow the lines of the proof of Lemma 11.2.4. We have $q(kG(K))$ is a four-dimensional Hopf subalgebra of B. Let ψ be the unique irreducible character of degree 6. Then we have $\psi^2 = \sum_{g \in G(K)} g + 4\psi$. This implies, as in the proof of Lemma 11.2.4, that $q(\psi) = \lambda_1 + \lambda_2 + \lambda_3$, where λ_i are pairwise distinct irreducible characters of degree 2 in B. Hence B must be of type $(1, 4; 2, 3)$ as a coalgebra. Hence, $q(kG(K)) = kG(B)$.

We know that $kG(B)$ is normal in B. Consider the sequence of surjective Hopf algebra maps $K \xrightarrow{q} B \xrightarrow{q'} B'$, where $B' = B/B(kG(B))^+$. Since $q(kG(K)) = kG(B)$, then we have $kG(K) \subseteq K^{\operatorname{co} q'q}$. Thus, $kG(K) = K^{\operatorname{co} q'q}$ is a normal Hopf subalgebra, as claimed.

Suppose next that $\dim A = 6$, so that $A = kS$, where S is a subgroup of order 6 in G. The quotient Hopf algebra K/KA^+ is cocommutative by Remark 3.2.7. Say $K/KA^+ = k\Gamma$, where $|\Gamma| = 8$. Then $K^* = k^\Gamma \#_\sigma k^S$ is a crossed product. Since there exists an irreducible K^*-module of dimension 6, then $S = M$ is abelian, by [**MW**, Proof of Theorem 2.1]. Thus K fits into the abelian extension $1 \to k^M \to K \to kN \to 1$, where N is a group of order 8 such that $K/KA^+ \simeq kN$. Dualizing, we get an abelian extension $1 \to k^N \to K^* \to kM \to 1$. Let $\triangleleft : N \times M \to N$, $\triangleright : N \times M \to M$, be the associated matched pair.

Since the action $\triangleright$ fixes $1 \in M$, and because N is of order 8, the set of fixed points M^N has at least 2 elements. Moreover, by the compatibility condition [**M**, (4.10)], M^N is a subgroup of M. It follows also from the formulas [**M**, (4.2) and (4.5)] for the multiplication and comultiplication of K that the subspace $B := k^N \otimes k(M^N)$ is a Hopf subalgebra of K^*. If $B = K^*$, then then the action $\triangleright$ is trivial. This implies that kM is a central Hopf subalgebra of K, which contradicts the assumption on the structure of $G(K)$.

Therefore $\dim B = 8|M^N| = 16$ or 24. If $\dim B = 16$, the lemma follows from the proof in the case $\dim A = 3$. If $\dim B = 24$, then $G(K) \cap Z(K) \neq 1$ and the lemma follows as well. □

CHAPTER 12

Dimension 54

12.1. First reduction

Let H be a nontrivial semisimple Hopf algebra of dimension 54.

LEMMA 12.1.1. *The order of $G(H^*)$ is either 2, 6, 9, 18 or 27 and as an algebra H is of one of the following types:*

- $(1, 2; 2, 4; 6, 1)$,
- $(1, 2; 2, 4; 3, 4)$,
- $(1, 2; 2, 13)$,
- $(1, 6; 2, 3; 6, 1)$,
- $(1, 6; 2, 12)$,
- $(1, 6; 2, 3; 3, 4)$,
- $(1, 9; 3, 1; 6, 1)$,
- $(1, 9; 3, 5)$,
- $(1, 18; 2, 9)$,
- $(1, 18; 3, 4)$,
- $(1, 18; 6, 1)$,
- $(1, 27; 3, 3)$.

PROOF. It follows from 1.1 and Corollary 2.2.3 that $|G(H^*)| \neq 1$. If $|G(H^*)| = 2$, then the possible algebra types for H other than the prescribed ones are

$$(1, 2; 2, 1; 4, 3), (1, 2; 3, 4; 4, 1), (1, 2; 4, 1; 6, 1),$$
$$(1, 2; 2, 9; 4, 1), (1, 2; 2, 5; 4, 2).$$

In the last two cases, by Theorems 2.2.1 and 2.4.2 H has a quotient Hopf algebra of algebra types $(1, 2; 2, 9)$ or $(1, 2; 2, 5)$, respectively, which contradicts [**NZ**]. In the third case, H has one irreducible character χ of degree 4, one irreducible character ψ of degree 6 and the remaining two are of degree 1. The character χ is necessarily stable under left multiplication by $G(H^*)$ and therefore we have a decomposition $\chi^2 = \chi\chi^* = \epsilon + g + n\chi + m\psi$, implying that $n = 2$ and $m = 1$. Then we have $1 = m = m(\chi, \psi\chi)$ and $\psi\chi = \chi + l\psi$, $l \in \mathbb{Z}$, which is not possible. The type $(1, 2; 2, 1; 4, 3)$ is discarded by Lemma 1.2.4.

Suppose that H is of type $(1,2;3,4;4,1)$. Write $G(H) = \{\epsilon, g\}$ and let τ be the unique irreducible character of degree 4. We must have $\tau\tau^* = 1 + g + 2\tau + \psi_1 + \psi_2$, where $\psi_1 \neq \psi_2$ are irreducible characters of degree 3. Then $m(\tau, \psi_1\tau) = 1$ and therefore, $\psi_1\tau = \tau + \sum_j \psi_j$, where ψ_j are irreducible of degree 3. Taking degrees we get a contradiction. This discards this possibility.

The possibility $|G(H^*)| = 3$ is discarded using Theorem 2.2.1 and 1.1. As to $|G(H^*)| = 6$, we need to discard the type $(1,6;4,3)$, which is done as follows: let χ be a irreducible character of degree 4, and write $\chi\chi^* = \sum_{g \in G[\chi]} g + \sum_i m_i\chi_i$, where χ_i are irreducible of degree 4; it turns out that 4 divides the order of $G[\chi]$ which is impossible.

Suppose that $|G(H^*)| = 9$. Then H has no irreducible characters of degree 2 by Theorem 2.2.1 and the only possibilities are the prescribed ones. The rest of the lemma follows also from 1.1. □

LEMMA 12.1.2. *Suppose that H is of type $(1,9;3,5)$ as a coalgebra. Then H is not simple.*

PROOF. We claim that there exist irreducible characters $\psi \neq \psi'$ of degree 3, which commute with $G(H)$, and such that

$$\psi\psi^* = \psi'(\psi')^* = \sum_{g \in G(H)} g.$$

Indeed, the group $G(H)$, being abelian, acts by left multiplication on the set $X_3' := \{\psi \in X_3 : g\psi = \psi g, \forall g \in G(H)\}$, and we have $|X_3'| = 2$ or 5. It follows that there are at least 2 stable elements in X_3', which therefore satisfy the claimed equations.

The claim implies that there are two subcoalgebras C_1 and C_2 of dimension 9, such that $gC_i = C_i = C_i g$, for all $g \in G(H)$. Moreover, we may assume that $k[C_1, C_2] = H$ since it is a Hopf subalgebra of dimension bigger than 18, which divides $\dim H$: if $\dim k[C_1, C_2] = 27$, then H is not simple by Corollary 1.4.3.

Fix $i = 1, 2$. By Proposition 3.2.6, $kG(H)$ is normal in $H = k[C_1, C_2]$. Therefore H is not simple in this case. □

LEMMA 12.1.3. *Suppose that H is of type $(1,9;3,1;6,1)$ as a coalgebra. Then H is not simple.*

PROOF. Let $D(H)$ be the Drinfeld double of H. By [**N**, Remark 2.2.4] $|GD(H)^*| \neq 1$. By [**R**], the group-like elements in $D(H)^*$ are of the form $g \otimes \eta$, where $g \in G(H)$, $\eta \in G(H^*)$, are such that $\eta \otimes g$ is central in $D(H)$. Observe that if the orders of g and η are different, say $n = |\eta| < |g|$, then the element $1 \otimes \epsilon \neq g^n \otimes \epsilon = (g \otimes \eta)^n$ would be such that $\epsilon \otimes g^n$ is central in $D(H)$, and *a fortiori*, $1 \neq g^n \in G(H) \cap Z(H)$.

Then we may assume that there is an element $g \otimes \eta \in GD(H)^*$, where $g \in G(H)$ and $\eta \in G(H^*)$ are of the same order. In particular, $|G(H^*)|$ is divisible by 3.

On the other hand, the irreducible characters of degree 1 and 3 span a standard subalgebra, which corresponds to a Hopf subalgebra A of dimension 18. Consider the projection $H^* \to A^*$; we may assume that $(H^*)^{\mathrm{co}\,A^*} = k1 \oplus V$, where V is an irreducible left coideal of dimension 2. Then, by Theorem 2.5.1, $|G(H^*)|$ is even and there is a quotient Hopf algebra $H \to B$, with $\dim B = 3|G(H^*)|$. Hence either $\dim B = 18$ and necessarily $H^{\mathrm{co}\,B} \subseteq kG(H)$, or else H^* is of type $(1,18;2,9)$ as a coalgebra and $H^{\mathrm{co}\,k^{G(H^*)}} \subseteq kG(H)$. This implies that H^* contains a normal Hopf subalgebra, and hence H is not simple. $\square$

REMARK 12.1.4. (i) If H is of type $(1,27;3,3)$, then H is not simple by Corollary 1.4.3.

(ii) Suppose that H is simple. It follows from Lemmas 12.1.1, 12.1.2 and 12.1.3 that there exist subgroups $F \subseteq G(H)$ and $F' \subseteq G(H^*)$ such that $|F| = |F'| = 2$. We have a projection $q : H \to k^{F'}$ such that $\dim H^{\mathrm{co}\,q} = 27$. Hence, $F \cap H^{\mathrm{co}\,q} = 1$ and $H = R \# kF$ is a biproduct. By Proposition 4.6.1, R is not commutative and not cocommutative.

(iii) Suppose that H is of type $(1,6;2,12)$ or $(1,18;2,9)$. Then H is not simple.

PROOF. Keep the notation in part (ii). We claim that $G(H)$ contains a central (hence unique) subgroup of order 2, which must stabilize all simple subcoalgebras of dimension 4. This implies, in view of Remark 3.2.7, that the braided Hopf algebra R is a cocommutative coalgebra and we are done.

To prove the claim we distinguish both cases. Suppose first that H is of type $(1,6;2,12)$. Then $G(H)$ is abelian by Proposition 1.2.6, and the claim follows in this case.

In the case of type $(1,18;2,9)$ we argue as follows. First note that by Lemma 2.3.2, $D(H)$ has an irreducible module of dimension 2. It follows from Theorem 2.2.1 that $G(D(H)^*)$ has an element $g \otimes \eta$ of order 2 or 3. If $g \otimes \eta$ is of order 2, we may assume that the center of $G(H)$ has an element of order 2, as claimed.

Finally suppose that $g \otimes \eta$ is of order 3. By [**N**, Corollary 2.3.2] there is an exact sequence of Hopf algebras $1 \to kG \to D(H) \to K \to 1$, and also $G(K^*) \neq 1$ by [**N**, Remark 2.2.4]; moreover, we may assume that $g \otimes \eta \in G(K^*) \subseteq G(D(H)^*)$. Therefore, by part (ii) in [**N**, Corollary 2.3.2], we get $\langle \eta, g \rangle = 1$.

Consider the natural projection $q : H \to k^{\langle \eta \rangle}$; we have shown that $\langle g \rangle \subseteq H^{\mathrm{co}\,q}$. Also, it is clear that $H^{\mathrm{co}\,q}$ contains every 2-subgroup of

$G(H)$. Since $H^{co\,q}$ is of dimension 18, and we may assume does not coincide with $kG(H)$, then $G(H) \cap H^{co\,q}$ is a normal subgroup of order 2 or 6 in $G(H)$. In particular, since we may assume that $1 \neq g \in Z(G(H))$, then the group $G(H) \cap H^{co\,q}$ is abelian. But $G(H) \cap H^{co\,q}$ contains a unique subgroup of order 2, then so does $G(H)$. This finishes the proof of the claim. □

(iv) By Theorem 2.2.1, since $\dim H$ is not divisible by 4, for every irreducible character χ of degree 2 we have $G[\chi] \neq 1$.

(v) If H is of type $(1, 2; 2, 4; 6, 1)$, $(1, 2; 2, 4; 3, 4)$, $(1, 6; 2, 3; 6, 1)$ or $(1, 6; 2, 3; 3, 4)$ as a coalgebra, then $G(H)$ and X_2 span a standard subalgebra of $R(H^*)$, which corresponds to a non-cocommutative Hopf subalgebra A of dimension 18. We may thus conclude that if H is simple, then H contains a Hopf subalgebra of dimension 18.

LEMMA 12.1.5. *Assume that H is of type $(1, 2; 2, 13)$ as a coalgebra. Then H is commutative.*

PROOF. By Theorem 4.6.5, we may assume that the order of $G(H^*)$ is odd; thus, by Lemma 12.1.1, $|G(H^*)| = 9$ or 27. Also, by Corollary 4.6.8, we may assume that $G(H) \cap Z(H) = 1$ and, in particular, $|G(H^*)| \neq 27$.

By [**N**, Remark 2.2.4], we have $G(D(H)^*) \neq 1$. Since $|G(H^*)|$ and $|G(H)|$ are relatively prime, the description of $G(D(H)^*)$ in [**R**] implies that there is an isomorphism

$$G(D(H)^*) \simeq (Z(H) \cap G(H)) \times (Z(H^*) \cap G(H^*)) = Z(H^*) \cap G(H^*).$$

Therefore, $n = |G(D(H)^*)| = 3$ or 9, and H^* contains a central group-like element of order 3. Hence H fits into an abelian cocentral extension

$$1 \to k^G \to H \to k\mathbb{Z}_3 \to 0,$$

where G is a group of order 18. Then the extension induces the trivial action $\triangleright : G \times \mathbb{Z}_3 \to \mathbb{Z}_3$, and an action by group automorphisms $\triangleleft : G \times \mathbb{Z}_3 \to G$. This implies that the transpose action $\rightharpoonup : k\mathbb{Z}_3 \otimes k^G \to k^G$ is by Hopf algebra automorphisms, and therefore that $kG(H) = k\widehat{G}$ is invariant under this action. Hence this action is trivial on $kG(H)$. By [**N**, Proposition 1.2.6] H is isomorphic as an algebra to the crossed product $H \simeq k^G \#_\rightharpoonup k\mathbb{Z}_3$. So it turns out that $kG(H)$ is central in H. This implies that H is commutative. □

LEMMA 12.1.6. *Assume that $|G(H)| = 18$. Then H is not simple.*

This discards the following possibilities for the coalgebra type of H:

$$(1, 18; 2, 9), \quad (1, 18; 3, 4), \quad (1, 18; 6, 1)$$

PROOF. In view of Remark 12.1.4 (iii) and Lemma 12.1.1, we may assume that H contains no irreducible left coideal of dimension 2. On the other hand, we know from Remark 12.1.4 (v) that there is a quotient Hopf algebra $H \to B$, where $\dim B = 18$. Then necessarily $H^{\operatorname{co} B} \subseteq kG(H)$ is a Hopf subalgebra, and thus H is not simple. $\square$

LEMMA 12.1.7. *Suppose that H is simple. Then H contains a Hopf subalgebra $A \subseteq H$, such that $A \simeq k^{\Gamma}$, where Γ is a non-abelian group of order 18. In particular, the dimension of an irreducible left H-module is at most 3.*

PROOF. By Remark 12.1.4 (v) and Lemma 12.1.6 we may assume that H contains a non-cocommutative Hopf subalgebra A of dimension 18. By [**M3**] there are two isomorphism classes of nontrivial semisimple Hopf algebras of dimension 18, which are dual to each another: $\mathcal{B}_0$ and $\mathcal{B}_1 = \mathcal{B}_0{}^*$. We have $|G(\mathcal{B}_0)| = 6$ and $|G(\mathcal{B}_1)| = 9$.

If A is not commutative, and since $G(A) \subseteq G(H)$, we may assume that $|G(A)| = 6$. By Remark 12.1.4 (ii), $H = R \# k\mathbb{Z}_2$, and since $k\mathbb{Z}_2 \subseteq A$ and A is not contained in $R = H^{\operatorname{co} \mathbb{Z}_2}$, then also $A \simeq R' \# k\mathbb{Z}_2$. In particular, the order of $G(A^*)$ is also divisible by 2, which implies that A is commutative as claimed. Finally, applying [**AN2**, Corollary 3.9] to the inclusion $k^{\Gamma} \subseteq H$, we find that $\dim V \le 3$, for all irreducible H-module V. $\square$

12.2. Main result

After what we have already shown until now, we can conclude that if H is simple, then the possible (co)algebra types for H are $(1, 2; 2, 4; 3, 4)$ and $(1, 6; 2, 3; 3, 4)$.

THEOREM 12.2.1. *Let H be a semisimple Hopf algebra of dimension 54 over k. Then H is not simple.*

PROOF. Let $A \subseteq H$ be the commutative Hopf subalgebra of dimension 18. Note that $G(H) \subseteq A$.

By Lemma 2.3.2 and Proposition 2.3.1, we have $G(D(H)^*) \neq 1$. Hence, we may assume that there exists a nontrivial one-dimensional Yetter-Drinfeld module for H. By Lemma 1.6.1, this has necessarily the form $V_{g,\eta}$, for some $1 \neq g \in G(H)$ and $1 \neq \eta \in G(H^*)$.

Consider the projection $q : H \to k\langle \eta \rangle$, obtained by transposing the inclusion $k\langle \eta \rangle \subseteq H^*$. Since A is commutative, $g^{-1}ag = a$, for all $a \in A^{\operatorname{co} q}$. By Theorem 1.6.4, we must have that $A^{\operatorname{co} q}$ is a Hopf subalgebra of A. But this is not possible since $\dim A^{\operatorname{co} q} = 9$, implying that $A^{\operatorname{co} q}$ is a cocommutative Hopf subalgebra, while $G(H)$ is of order 2 or 6. This contradiction finishes the proof of the theorem. $\square$

CHAPTER 13

Dimension 56

13.1. First reduction

Let H be a nontrivial semisimple Hopf algebra of dimension 56.

LEMMA 13.1.1. *The order of $G(H^*)$ is either 4, 7, 8 or 28, and as an algebra H is of one of the following types:*

- $(1, 4; 2, 13)$,
- $(1, 4; 2, 9; 4, 1)$,
- $(1, 4; 2, 5; 4, 2)$,
- $(1, 4; 2, 1; 4, 3)$,
- $(1, 7; 7, 1)$,
- $(1, 8; 4, 3)$,
- $(1, 8; 2, 4; 4, 2)$,
- $(1, 8; 2, 8; 4, 1)$,
- $(1, 8; 2, 12)$,
- $(1, 28; 2, 7)$.

PROOF. We have $G(H^*) \neq 1$ by Corollary 2.2.3; indeed, counting arguments show that the assumption $G(H^*) = 1$ implies that H must have an irreducible character of degree 2. Suppose that $|G(H^*)| = 2$, and H is of type $(1, 2; 2, n_2; 3, n_3; \dots)$. If $n_2 \neq 0$, by Theorems 2.2.1 and 2.4.2, there is a quotient Hopf algebra of dimension $2(2n_2+1)$; thus $n_2 = 3$ by [**NZ**], and using 1.1 we find a contradiction. The possibilities with $n_2 = 0$ are the types $(1, 2; 3, 2; 6, 1)$ and $(1, 2; 3, 6)$: these cases are seen to be impossible, after trying to decompose the product $\psi\psi^*$ into irreducibles, where ψ is an irreducible character of degree 3.

In the case $|G(H^*)| = 4$, apart from the types listed in the lemma, we have the following possibilities:

$$(1, 4; 2, 4; 6, 1), (1, 4; 2, 4; 3, 4), (1, 4; 3, 4; 4, 1), (1, 4; 4, 1; 6, 1).$$

In the first two cases, H has a quotient Hopf algebra of dimension 20, which is impossible. In the third case, H has four degree 1 representations, four irreducible characters $\chi_1, \dots, \chi_4$ of degree 3, and one irreducible character ψ of degree 4. Then we must have $m(\chi_i, \psi^2) = m(\chi_i, \psi\psi^*) > 0$ for some i; also, since $G(H^*)$ must permute transitively the degree 3 characters under left or right multiplication and since $g\psi = \psi$, for all $g \in G(H^*)$, then $m(\chi_i, \psi^2) > 0$ for all $i = 1, \ldots, 4$, implying that $m(\psi, \chi_i\psi) = m(\chi_i, \psi^2) = 1$ for all $i = 1, \ldots, 4$. Thus $\chi_1\psi = \psi + \sum_i m_i\chi_i$, where m_i are non-negative integers not all them equal to zero. Again since $\psi g = \psi$ for all g, we find that $m_i \neq 0$, for all $i = 1, \ldots, 4$. Taking degrees we find a contradiction. This discards the type $(1, 4; 3, 4; 4, 1)$.

The type $(1, 4; 4, 1; 6, 1)$ is discarded as follows: let χ and ψ be the unique irreducible characters of degrees 4 and 6, respectively. Write $\chi^2 = \sum_{g \in G(H^*)} g + n\chi + m\psi$; thus $m \neq 0$, since this would give a quotient Hopf algebra of dimension 20, which is not possible. Hence $m = 2$, and since $m = m(\chi, \psi\chi)$, we have $\psi\chi = 2\chi + l\psi$. Taking degrees we get a contradiction.

It is easy to see that $|G(H^*)| = 14$ is not possible. The rest of the lemma follows from 1.1. □

REMARK 13.1.2. (i) If H is of type $(1, 28; 2, 7)$, then H is not simple, by Corollary 1.4.3.

(ii) It follows from Proposition 2.1.3 that, except for the coalgebra type $(1, 7; 7, 1)$, H has a Hopf subalgebra of dimension 8.

(iii) By [**N2**], if $G(H) \cap Z(H) = 1 = G(H^*) \cap Z(H^*)$, then $G(H)$ and $G(H^*)$ are not both of order 8.

LEMMA 13.1.3. *Suppose that H is of type $(1, 7; 7, 1)$ as a coalgebra. Then H is not simple.*

PROOF. We may assume that H^* is also of type $(1, 7; 7, 1)$ as a coalgebra; otherwise, there is a quotient Hopf algebra $H \to B$, where $\dim B = 8$ and necessarily $H^{\operatorname{co} B} = kG(H)$ is a normal Hopf subalgebra. Hence $H = R \# kG(H)$ is a biproduct, where $\dim R = 8$ and moreover, by Remark 3.2.7, R is cocommutative. Then the lemma follows from Proposition 4.6.1. □

LEMMA 13.1.4. *Suppose that H is simple.*

Then $H = R \# A$, where A is a semisimple Hopf algebra of dimension 8 and R is a Yetter-Drinfeld Hopf algebra over A of dimension 7.

PROOF. We may assume that H has a Hopf subalgebra A and a Hopf algebra quotient $q : H \to B$, such that $\dim A = \dim B = 8$.

Since $\dim H^{\operatorname{co} B} = 7$, we may assume that $kG(H) \cap H^{\operatorname{co} B} = k1$. In particular, the lemma follows in the case where A is cocommutative.

If A is not cocommutative, then A contains a unique 4-dimensional simple coalgebra C of H. Note that C is not contained in $H^{\operatorname{co} B}$ since $G(H) \subseteq C^2$.

If the restriction $q|_A : A \to B$ is an isomorphism, then we are done. Otherwise, $H^{co\,B}$ contains a 2-dimensional irreducible left coideal V of A. But since $H^{co\,B}$ cannot contain C, we must have $H^{co\,B} \cap A = k1 \oplus V$. This contradicts Lemma 1.3.4. The proof of the lemma is complete. $\square$

13.2. Main result

In this section, we apply Lemma 13.1.4 to show that a semisimple Hopf algebra of dimension 56 is not simple as a Hopf algebra.

THEOREM 13.2.1. *Let H be a semisimple Hopf algebra of dimension 56. Then H is not simple.*

PROOF. Suppose on the contrary that H is simple. Keep the notation in Lemma 13.1.4. After dualizing if necessary, we may assume that $|G(H)| = 4$ (see Remark 13.1.2 (ii)); so that A is not cocommutative. As a left coideal of H, R must be of one of the following types:

$$(1)\ k1 \oplus V_1 \oplus V_2 \oplus V_3, \quad (2)\ k1 \oplus V \oplus W,$$

where $\dim V_j = 2 = \dim V$, for all $j = 1, 2, 3$, and $\dim W = 4$. In particular, the type $(1, 4; 2, 1; 4, 3)$ is impossible.

Consider first the case (1). By Lemma 4.3.3 $\rho(V_j) \subseteq kG(A) \otimes V_j$, for all j; thus $\rho(R) \subseteq kG(A) \otimes R$ and therefore $R\#kG(A)$ is a Hopf subalgebra of H. This implies that H is not simple, since $G(A)$ has index 2 in A.

Suppose finally that R is as in case (2). By Lemma 4.3.3 $\rho(V) \subseteq kG(A) \otimes V$. Moreover, $kG(A).V \subseteq V$, since gVg^{-1} is a 2-dimensional irreducible left coideal of H contained in R, for all $g \in G(H)$.

By Proposition 4.2.1, V is an A-subcomodule subcoalgebra of R. Then, by Lemma 4.3.1, $B = k[V]\#kG(A)$ is a Hopf subalgebra of H. Since $\dim B$ is divisible by 4 and $\dim k[V] \geq 3$, we must have $\dim B = 28$ and $kV = R$. As before, we get that $R\#kG(A)$ is a Hopf subalgebra of index 2 in H, and H is not simple. $\square$

APPENDIX A

Drinfeld Double of H_8

We shall denote by H_8 the unique nontrivial 8-dimensional semisimple Hopf algebra over k [**KP, M4**]. We shall use the notation D_4 and Q to indicate, respectively, the dihedral and quaternionic groups of order 8. For a Hopf algebra A, $D(A)$ denotes the Drinfeld double of A.

By [**M4**] the only non-commutative semisimple Hopf algebras of dimension 8 are H_8, kD_4 and kQ.

A.1. Structure of $D(H_8)$

Tambara and Yamagami show in [**TY**] that the categories of representations of these three Hopf algebras are not equivalent as monoidal categories. The comparison of Schur indicators implies that the representation theory of H_8 is in some sense closer to that of kD than to that of kQ; see [**Mo3**]. On the other hand, note that H_8 fits into an exact sequence $1 \to k^\Gamma \to H_8 \to kF \to 1$, where $\Gamma \simeq \mathbb{Z}_2 \times \mathbb{Z}_2$ and $F \simeq \mathbb{Z}_2$, such that the associated product group associated to the extension is D_4.

The main results in this appendix are the following:

THEOREM A.1.1. *(i)* $D(H_8)$ *fits into an abelian central extension*

$$0 \to k^G \to D(H_8) \to kG \to 1,$$

where $G = G(D(H_8)^*) \simeq \mathbb{Z}_2 \times \mathbb{Z}_2 \times \mathbb{Z}_2$.

(ii) $D(H_8)$ *is of type* $(1, 8; 2, 14)$ *as an algebra, and as a coalgebra it is of type* $(1, 16; 2, 8; 4, 1)$.

THEOREM A.1.2. $D(H_8)$ *has no quotient Hopf algebra isomorphic to* kQ.

Observe that since H_8 admits quasitriangular structures [**Su**], then $D(H_8)$ also has quotient Hopf algebras isomorphic to H_8.

REMARK A.1.3. Let G be a finite group. Then $D(G)$ is a semidirect product $D(G) = k^G \rtimes kG$, with respect to the action comming from the adjoint action of G on itself. Hence, the irreducible representations of $D(G)$, viewed as irreducible Yetter-Drinfeld modules over kG, are classified by the modules $V_{g,\rho} := kG \otimes_{Z_G(g)} \rho$, where g runs over a system of representatives of the conjugacy classes in G and ρ runs over the irreducible representations of the centralizer $Z_G(g)$.

This implies that if $|G| = p^3$, p prime, then $D(G)$ has exactly p^3 one-dimensional representations and the remaining irreducible representations are of dimension p; that is, $D(G)$ is of type $(1, p^3; p, p(p^3-1))$ as an algebra.

In particular, the Drinfeld doubles of the three non-commutative 8-dimensional semisimple Hopf algebras have the same algebra structure.

A.2. Proof of Theorem A.1.1

Let A be a finite-dimensional Hopf algebra and let $g \in G(A)$, $\eta \in G(A^*)$. Let $V_{g,\eta}$ denote the vector space $k1$ endowed with the action $h.1 = \eta(h)1$, $h \in H$, and the coaction $1 \mapsto g \otimes 1$.

By Lemma 1.6.1, the one-dimensional Yetter-Drinfeld modules over A are exactly of the form $V_{g,\eta}$, where $g \in G(A)$ and $\eta \in G(A^*)$ are such that $(\eta \rightharpoonup h)g = g(h \leftharpoonup \eta)$, for all $h \in A$.

Note that if the condition $(\eta \rightharpoonup h)g = g(h \leftharpoonup \eta)$ holds for all h in a set of generators of A, then it holds for all $h \in A$.

It turns out that the elements of the form $\eta \otimes g$, where g and η satisfy the condition in Lemma 1.6.1 are exactly the central group-like elements in $D(A)$ [**R**]. In particular, $V_{g,\epsilon}$ (respectively, $V_{1,\eta}$) is a Yetter-Drinfeld module if and only if $g \in Z(A)$ (respectively, $\eta \in Z(A^*)$).

As in [**M4**], H_8 can be presented by generators x, y, z with relations

$$x^2 = y^2 = 1,$$
$$xy = yx, \quad zx = yz, \quad zy = xz,$$
$$z^2 = \frac{1}{2}(1 + x + y - xy).$$

The coalgebra structure is determined by

$$\Delta(x) = x \otimes x, \qquad \Delta(y) = y \otimes y,$$
$$\Delta(z) = \frac{1}{2}\left((1+y) \otimes 1 + (1-y) \otimes x\right)(z \otimes z).$$

In particular, $z \in H_8^\times$, $\epsilon(z) = 1$ and $\mathcal{S}(z) = z^{-1}$. We have in addition $H_8 \simeq H_8^*$, $G(H_8) = \{1, x, y, xy\} \simeq \mathbb{Z}_2 \times \mathbb{Z}_2$, and $G(H_8) \cap Z(H_8) = \{1, xy\}$.

LEMMA A.2.1. *Let $g \in G(H_8) \backslash Z(H_8)$ and $\eta \in G(H_8^*) \backslash Z(H_8^*)$. Then $V_{g,\eta}$ is a Yetter-Drinfeld module of H_8.*

This implies that $D(H_8)$ has exactly 8 distinct one-dimensional representations. In other words, we have

$$(A.2.2) \qquad G(D(H_8)^*) = \langle xy \otimes \epsilon \rangle \oplus \langle 1 \otimes \alpha\beta \rangle \oplus \langle x \otimes \alpha \rangle,$$

where $G(H_8^*) = \{\epsilon, \alpha, \beta, \alpha\beta\}$ and $G(H_8^*) \cap Z(H_8^*) = \{\epsilon, \alpha\beta\}$.

PROOF. We shall show that if $1, xy \neq g$, then every element $\eta \in G(H_8^*) \backslash Z(H_8^*)$ satisfies the condition in Lemma 1.6.1.

Notice that for any one-dimensional representation $\eta : H_8 \to k$ we must have $\eta(x) = \eta(y)$, because of the relation $zx = yz$ and the fact that $z \in H_8^\times$. Moreover, if η is not central in H_8^*, then $\eta(x) = \eta(y) = -1$: indeed, if $\eta|_{G(H)} = 1$, then $k\langle \eta \rangle = (H_8^*)^{\operatorname{co} G(H_8)}$ and thus $\eta \in Z(H_8^*)$.

Also, it is enough to see that the condition in Lemma 1.6.1 is satisfied for $h = z$, since it is always satisfied for $h = x, y$ and these generate H_8 as an algebra.

We compute

$$(A.2.3) \qquad \eta \rightharpoonup z = \eta(z_2)z_1 = \frac{1}{2}\left(1 + y + \eta(x)(1-y)\right)\eta(z)z,$$

$$(A.2.4) \qquad z \leftharpoonup \eta = \eta(z_1)z_2 = \frac{1}{2}\left(\eta(1+y)1 + \eta(1-y)x\right)\eta(z)z.$$

Replacing in this identities $g = y$, the condition $(\eta \rightharpoonup z)y = y(z \leftharpoonup \eta)$ is equivalent to the equation $xy + x + \eta(x)(x - xy) = \eta(1+y)y + \eta(1-y)xy$. This is always satisfied for $\eta \in G(H_8^*) \backslash Z(H_8^*)$.

The argument for $g = x$ is similar. $\square$

COROLLARY A.2.5. $G(D(H_8)^*) \simeq \mathbb{Z}_2 \times \mathbb{Z}_2 \times \mathbb{Z}_2$.

PROOF. We have $|G(D(H_8)^*)| = 8$ and $G(D(H_8)^*)$ is isomorphic to a subgroup of $G(D(H_8)) = G(H_8^*) \times G(H_8)$. $\square$

Proof of Theorem A.1.1. (i) In virtue of [**N**, Corollary 2.3.2] and Corollary A.2.5, there is a central extension

$$0 \longrightarrow kG \xrightarrow{\iota} D(H_8) \xrightarrow{\pi} K \longrightarrow 1,$$

where $G \simeq G(D(H_8)^*) \simeq \mathbb{Z}_2 \times \mathbb{Z}_2 \times \mathbb{Z}_2$, and the map ι is determined by $\iota(g \otimes \eta) = \eta \otimes g$. In particular, $\dim K = 8$.

We have $kG \simeq k^G$. Identify K with a Hopf subalgebra of $D(H_8)^*$. Part (i) will be established if we show that $G(K^*) = G$. To see this, we observe that a group-like element $g \otimes \eta$ belongs to $G(K^*)$ exactly when it is a one dimensional representation of $D(H_8)$ which factorizes through K; that is, $g \otimes \eta$ belongs to $G(K^*)$ if and only if

$$(A.2.6) \qquad \langle \eta, g' \rangle \langle \eta', g \rangle = 1,$$

for all group-like elements $g' \otimes \eta' \in G(D(H_8)^*)$. See [**N**, Corollary 2.3.2]. Finally, using the description of the elements in $G(D(H_8)^*)$ given in (A.2.2), one sees that $G(K^*) = G$, thus proving part (i).

(ii) As a coalgebra, $D(H_8)$ is a tensor product: $D(H_8) = (H_8^*)^{\mathrm{cop}} \otimes H_8$. This proves the statement corresponding to the coalgebra structure.

Combining part (i) with the description in [**KMM**, Theorem 3.3] (the action being trivial in our situation), we find that the simple $D(H_8)$-modules are of the form $p_g \otimes V$, where $g \in G$ and V is an irreducible $k_{\sigma_g}G$-irreducible module, for some 2-cocycle $\sigma_g : G \times G \to k^\times$. The dimensions of the irreducible $k_{\sigma_g}G$-modules are either 1 or 2 (since they divide the order of G and their square is less than 8). This finishes the proof of (ii). $\square$

A.3. Proof of Theorem A.1.2

Suppose that $\pi : D(H_8) \to kM$ is a surjective Hopf algebra map, where M is a group of order 8. Then $kM = \pi(H_8^*)\pi(H_8)$ and $\dim \pi(H_8^*)$ and $\dim \pi(H_8)$ divide 4 (because $H_8 \simeq H_8^*$ is not isomorphic to kM). Then some of them, say $\pi(H_8)$ is of dimension 4. Then $\pi(H_8) \simeq \mathbb{Z}_2 \times \mathbb{Z}_2$, since H_8^* has no Hopf subalgebra isomorphic to $\mathbb{Z}_4$. Therefore $M \neq Q$. $\square$

Bibliography

[A] N. Andruskiewitsch, *Notes on extensions of Hopf algebras*, Canad. J. Math. **48** (1996), 3-42.

[A2] N. Andruskiewitsch, *About finite dimensional Hopf algebras*, Contemp. Math. **294** (2002), 1–57.

[AN] N. Andruskiewitsch and S. Natale, *Examples of self-dual Hopf algebras*, J. Math. Sci. Univ. Tokyo **6** (1999), 181-215.

[AN2] N. Andruskiewitsch and S. Natale, *Harmonic analysis on semisimple Hopf algebras*, Algebra i Analiz **12** (2000), 3–27.

[AN3] N. Andruskiewitsch and S. Natale, *Braided Hopf algebras arising from matched pairs of groups*, J. Pure Appl. Algebra **132** (2003), 119–149.

[CR] C. Curtis and I. Reiner, *Methods of Representation Theory* Vol. I, Wiley Interscience, New York, 1990.

[EG] P. Etingof and S. Gelaki, *Semisimple Hopf algebras of dimension pq are trivial*, J. Algebra **210** (1998), 664-669.

[EG2] P. Etingof and S. Gelaki, *Some properties of finite dimensional semisimple Hopf algebras*, Math. Res. Lett. **5** (1998), 551-561.

[EG3] P. Etingof and S. Gelaki, *The representation theory of co-triangular semisimple Hopf algebras*, Int. Math. Res. Not. **7** (1999), 387–394.

[EG4] P. Etingof and S. Gelaki, *The classification of triangular semisimple and cosemisimple Hopf algebras over an algebraically closed field*, Int. Math. Res. Not. **5** (2000), 223–234.

[EG5] P. Etingof and S. Gelaki, *Isocategorical groups*, Int. Math. Res. Not. **2** (2001), 59–76.

[F] N. Fukuda, *Semisimple Hopf algebras of dimension* 12, Tsukuba J. Math. **21** (1997), 43–54.

[GN] C. Galindo and S. Natale, *Simple Hopf algebras and deformations of finite groups*, preprint `math.QA/0608734` (2006).

[G] S. Gelaki, *Quantum groups of dimension pq^2*, Israel J. Math. **102** (1997), 227–267.

[GW] S. Gelaki and S. Westreich, *On semisimple Hopf algebras of dimension pq*, Proc. Am. Math. Soc. **128** (2000), 39–47. (Corrigendum: Proc. Am. Math. Soc. **128** (2000), 2829–2831.)

[IK] M. Izumi and H. Kosaki, *Kac algebras arising from composition of subfactors: general theory and classification*, Mem. Amer. Math. Soc. **158**, 750, (2002).

[Ka] G. Kac, *Extensions of groups to ring groups*, Math. USSR. Sb. **5** (1968), 451–474.

[Ka2] G. Kac, *Certain arithmetic properties of ring groups*, Functional Anal. Appl. **6** (1972), 158–160.

[KP] G. Kac and V. Paljutkin, *Finite ring groups*, Trudy Moskov. Mat. Obš č. **15** (1966) 224–261.

[K] Y. Kashina, *Classification of semisimple Hopf algebras of dimension* 16, J. Algebra **232** (2000), 617–663.

[K2] Y. Kashina, *Some results on Hopf algebras of Frobenius type*, Math. Appl., Dordr. **555** (2003), 85-104.

[KMM] Y. Kashina, G. Mason and S. Montgomery, *Computing the Frobenius-Schur indicator for abelian extensions of Hopf algebras*, J. Algebra **251** (2002), 888–913.

[KM] T. Kobayashi and A. Masuoka, *A result extended from groups to Hopf algebras*, Commun. Algebra **25** (1997), 1169–1197.

[LR] R. Larson and D. Radford, *Semisimple cosemisimple Hopf algebras*, Amer. J. Math. **110** (1988), 187–195.

[LR2] R. Larson and D. Radford, *Finite dimensional cosemisimple Hopf algebras in characteristic* 0 *are semisimple*, J. Algebra **117** (1988), 267–289.

[Mj] S. Majid, *Crossed products by braided groups and bosonization*, J. Algebra **163** (1994), 165–190.

[Mj2] S. Majid, *Foundations of Quantum Group Theory*, Cambridge Univ. Press (1995).

[M] A. Masuoka, *Extensions of Hopf algebras*, Trabajos de Matemática 41/99, FaMAF (1999).

[M2] A. Masuoka, *Freeness of Hopf algebras over coideal subalgebras*, Commun. Algebra **20** (1992) 1353-1373.

[M3] A. Masuoka, *Some further classification results on semisimple Hopf algebras*, Commun. Algebra **24** (1996), 307–329.

[M4] A. Masuoka, *Semisimple Hopf algebras of dimension* 6, 8, Israel J. Math. **92** (1995), 361–373.

[M5] A. Masuoka, *Cocycle deformations and Galois objects for some co-semisimple Hopf algebras of finite dimension*, Contemp. Math. **267** (2000), 195–214.

[M6] A. Masuoka, *The p^n-th Theorem for Hopf algebras*, Proc. Amer. Math. Soc. **124** (1996), 187–195.

[M7] A. Masuoka, *Self dual Hopf algebras of dimension p^3 obtained by extension*, J. Algebra **178** (1995), 791–806.

[M8] A. Masuoka, *Semisimple Hopf algebras of dimension* $2p$, Comm. Algebra **23** (1995), 1931–1940.

[M9] A. Masuoka, *Coalgebras actions on Azumaya algebras*, Tsukuba J. Math. **14** (1990), 107–112.

[M10] A. Masuoka, *Hopf algebra extensions and cohomology*, Math. Sci. Res. Inst. Publ. **43** (2002), 167–209.

[Mo] S. Montgomery, *Hopf Algebras and their Actions on Rings*, CMBS Reg. Conf. Ser. in Math. **82**, Amer. Math. Soc., 1993.

[Mo1] S. Montgomery, *Classifying finite dimensional semisimple Hopf algebras*, Contemp. Math. **229** (1998), 265–279.

[Mo2] S. Montgomery, *Indecomposable coalgebras, simple comodules and poin-ted Hopf algebras*, Proc. Amer. Math. Soc. **123** (1995), 2343–2351.

[Mo3] S. Montgomery, *Representation theory of semisimple Hopf algebras*, Algebra—representation theory (Constanta, 2000), 189–218, NATO Sci. Ser. II Math. Phys. Chem. **28**, Kluwer Acad. Publ., Dordrecht, 2001.

[MW] S. Montgomery and S. Whiterspoon, *Irreducible representations of crossed products*, J. Pure Appl. Algebra **129** (1998), 315–326.

[N] S. Natale, *On semisimple Hopf algebras of dimension pq^2*, J. Algebra **221** (1999), 242–278.

[N2] S. Natale, *On semisimple Hopf algebras of dimension pq^2, II*, Algebr. Represent. Theory **5** (3), (2001), 277–291.

[N3] S. Natale, *On semisimple Hopf algebras of dimension pq^r*, Algebr. Represent. Theory **7** (2) (2004), 173-188.

[N4] S. Natale, *On semisimple Hopf algebras of low dimension*, AMA Algebra Montp. Announc. **01** (2003).

[N5] S. Natale, *On group theoretical Hopf algebras and exact factorizations of finite groups*, J. Algebra **270** (1) (2003), 199-211. Preprint math.QA/0208054.

[NR] W. Nichols and M. Richmond, *The Grothendieck group of a Hopf algebra*, J. Pure Appl. Algebra **106** (1996), 297–306.

[NZ] W. Nichols and M. Zoeller, *A Hopf algebra freeness Theorem*, Amer. J. Math. **111** (1989), 381–385.

[Nk] D. Nikshych, *K_0-rings and twisting of finite-dimensional semisimple Hopf algebras*, Commun. Algebra **26** (1998), 321–342. (Corrigendum: Commun. Algebra **26** (1998), 2019.)

[R] D. Radford, *Minimal quasitriangular Hopf algebras*, J. Algebra **157** (1993), 285–315.

[R2] D. Radford, *The structure of Hopf algebras with a projection*, J. Algebra **92** (1985), 322–347.

[Sb] P. Schauenburg, *On the braiding on a Hopf algebra in a braided category*, New York J. Math. **4** (1998), 259-263.

[Sc] H.-J. Schneider, *Lectures on Hopf algebras*, Trabajos de Matemática 31/95, FaMAF (1995).

[Sc2] H.-J. Schneider, *Normal basis and transitivity of crossed products for Hopf algebras*, J. Algebra **152** (1992), 289–312.

[Sc3] H.-J. Schneider, *Principal homogeneous spaces for arbitrary Hopf algebras*, Israel J. Math. **72** (1990), 167–195.

[So] Y. Sommerhäuser, *Yetter-Drinfel'd Hopf algebras over groups of prime order*, Lecture Notes in Math. **1789** (2002), Springer-Verlag.

[Su] S. Suzuki, *A family of braided cosemisimple Hopf algebras of finite dimension*, Tsukuba J. Math. **22** (1998), 1-29.

[T] D. Tambara, *Representation of tensor categories with fusion rules of self-duality for finite abelian groups*, Israel J. Math. **118** (2000), 29–60.

[TY] D. Tambara and S. Yamagami, *Tensor categories with fusion rules of self-duality for finite abelian groups*, J. Algebra **209** (1998), 29–60.

[ZS] S. Zhu, *On finite dimensional Hopf algebras*, Commun. Algebra **21** (1993), 3871–3885.

[Z] Y. Zhu, *Hopf algebras of prime dimension*, Int. Math. Res. Not. **1** (1994), 53–59.

[Z2] Y. Zhu, *A commuting pair in Hopf algebras*, Proc. Amer. Math. Soc. **125** (1997), 2847–2851.

Editorial Information

To be published in the *Memoirs*, a paper must be correct, new, nontrivial, and significant. Further, it must be well written and of interest to a substantial number of mathematicians. Piecemeal results, such as an inconclusive step toward an unproved major theorem or a minor variation on a known result, are in general not acceptable for publication.

Papers appearing in *Memoirs* are generally at least 80 and not more than 200 published pages in length. Papers less than 80 or more than 200 published pages require the approval of the Managing Editor of the Transactions/Memoirs Editorial Board.

As of November 30, 2006, the backlog for this journal was approximately 12 volumes. This estimate is the result of dividing the number of manuscripts for this journal in the Providence office that have not yet gone to the printer on the above date by the average number of monographs per volume over the previous twelve months, reduced by the number of volumes published in four months (the time necessary for preparing a volume for the printer). (There are 6 volumes per year, each usually containing at least 4 numbers.)

A Consent to Publish and Copyright Agreement is required before a paper will be published in the *Memoirs*. After a paper is accepted for publication, the Providence office will send a Consent to Publish and Copyright Agreement to all authors of the paper. By submitting a paper to the *Memoirs*, authors certify that the results have not been submitted to nor are they under consideration for publication by another journal, conference proceedings, or similar publication.

Information for Authors

Memoirs are printed from camera copy fully prepared by the author. This means that the finished book will look exactly like the copy submitted.

Initial submission. The AMS uses Centralized Manuscript Processing for initial submissions. Authors should submit a PDF file using the Initial Manuscript Submission form found at www.ams.org/cgi-bin/peertrack/submission.pl, or send one copy of the manuscript to the following address: Centralized Manuscript Processing, MEMOIRS OF THE AMS, 201 Charles Street, Providence, RI 02904-2294 USA. If a paper copy is being forwarded to the AMS, indicate that it is for it Memoirs and include the name of the corresponding author, contact information such as email address or mailing address, and the name of an appropriate Editor to review the paper (see the list of Editors below).

The paper must contain a *descriptive title* and an *abstract* that summarizes the article in language suitable for workers in the general field (algebra, analysis, etc.). The *descriptive title* should be short, but informative; useless or vague phrases such as "some remarks about" or "concerning" should be avoided. The *abstract* should be at least one complete sentence, and at most 300 words. Included with the footnotes to the paper should be the 2000 *Mathematics Subject Classification* representing the primary and secondary subjects of the article. The classifications are accessible from www.ams.org/msc/. The list of classifications is also available in print starting with the 1999 annual index of *Mathematical Reviews*. The Mathematics Subject Classification footnote may be followed by a list of *key words and phrases* describing the subject matter of the article and taken from it. Journal abbreviations used in bibliographies are listed in the latest *Mathematical Reviews* annual index. The series abbreviations are also accessible from www.ams.org/publications/. To help in preparing and verifying references, the AMS offers MR Lookup, a Reference Tool for Linking, at www.ams.org/mrlookup/.

Electronically prepared manuscripts. The AMS encourages electronically prepared manuscripts, with a strong preference for $\mathcal{AMS}$-LaTeX. To this end, the Society has prepared $\mathcal{AMS}$-LaTeX author packages for each AMS publication. Author packages include instructions for preparing electronic manuscripts, samples, and a style file that generates the particular design specifications of that publication series. Though $\mathcal{AMS}$-LaTeX is the highly preferred format of TeX, author packages are also available in $\mathcal{AMS}$-TeX.

Authors may retrieve an author package from the AMS website starting from `www.ams.org/tex/` or via FTP to `ftp.ams.org` (login as `anonymous`, enter username as password, and type `cd pub/author-info`). The *AMS Author Handbook* and the *Instruction Manual* are available in PDF format following the author packages link from `www.ams.org/tex/`. The author package can also be obtained free of charge by sending email to `tech-support@ams.org` (Internet) or from the Publication Division, American Mathematical Society, 201 Charles St., Providence, RI 02904-2294, USA. When requesting an author package, please specify $\mathcal{AMS}$-LaTeX or $\mathcal{AMS}$-TeX and the publication in which your paper will appear. Please be sure to include your complete mailing address.

After acceptance. The final version of the electronic file should be sent to the Providence office (this includes any TeX source file, any graphics files, and the DVI or PostScript file) immediately after the paper has been accepted for publication.

Before sending the source file, be sure you have proofread your paper carefully. The files you send must be the EXACT files used to generate the proof copy that was accepted for publication. For all publications, authors are required to send a printed copy of their paper, which exactly matches the copy approved for publication, along with any graphics that will appear in the paper.

Accepted electronically prepared files can be submitted via the web at `www.ams.org/submit-book-journal/`, sent via FTP, or sent on CD-Rom or diskette to the Electronic Prepress Department, American Mathematical Society, 201 Charles Street, Providence, RI 02904-2294 USA. TeX source files, DVI files, and PostScript files can be transferred over the Internet by FTP to the Internet node `ftp.ams.org` (130.44.1.100). When sending a manuscript electronically via CD-Rom or diskette, please be sure to include a message identifying the paper as a Memoir.

Electronically prepared manuscripts can also be sent via email to `pub-submit@ams.org` (Internet). In order to send files via email, they must be encoded properly. (DVI files are binary and PostScript files tend to be very large.)

Electronic graphics. Comprehensive instructions on preparing graphics are available at `www.ams.org/jourhtml/`. A few of the major requirements are given here.

Submit files for graphics as EPS (Encapsulated PostScript) files. This includes graphics originated via a graphics application as well as scanned photographs or other computer-generated images. If this is not possible, TIFF files are acceptable as long as they can be opened in Adobe Photoshop or Illustrator. No matter what method was used to produce the graphic, it is necessary to provide a paper copy to the AMS.

Authors using graphics packages for the creation of electronic art should also avoid the use of any lines thinner than 0.5 points in width. Many graphics packages allow the user to specify a "hairline" for a very thin line. Hairlines often look acceptable when proofed on a typical laser printer. However, when produced on a high-resolution laser imagesetter, hairlines become nearly invisible and will be lost entirely in the final printing process.

Screens should be set to values between 15% and 85%. Screens which fall outside of this range are too light or too dark to print correctly. Variations of screens within a graphic should be no less than 10%.

Inquiries. Any inquiries concerning a paper that has been accepted for publication should be sent to `memo-query@ams.org` or directly to the Electronic Prepress Department, American Mathematical Society, 201 Charles St., Providence, RI 02904-2294 USA.

Editors

This journal is designed particularly for long research papers, normally at least 80 pages in length, and groups of cognate papers in pure and applied mathematics. Papers intended for publication in the *Memoirs* should be addressed to one of the following editors. The AMS uses Centralized Manuscript Processing for initial submissions to AMS journals. Authors should follow instructions listed on the Initial Submission page found at www.ams.org/memo/memosubmit.html.

Algebra to ALEXANDER KLESHCHEV, Department of Mathematics, University of Oregon, Eugene, OR 97403-1222; email: ams@noether.uoregon.edu

Algebra and its application to MINA TEICHER, Emmy Noether Research Institute for Mathematics, Bar-Ilan University, Ramat-Gan 52900, Israel; email: teicher@macs.biu.ac.il

Algebraic geometry to DAN ABRAMOVICH, Department of Mathematics, Brown University, Box 1917, Providence, RI 02912; email: amsedit@math.brown.edu

Algebraic number theory to V. KUMAR MURTY, Department of Mathematics, University of Toronto, 100 St. George Street, Toronto, ON M5S 1A1, Canada; email: murty@math.toronto.edu

Algebraic topology to ALEJANDRO ADEM, Department of Mathematics, University of British Columbia, Room 121, 1984 Mathematics Road, Vancouver, British Columbia, Canada V6T 1Z2; email: adem@math.ubc.ca

Combinatorics to JOHN R. STEMBRIDGE, Department of Mathematics, University of Michigan, Ann Arbor, Michigan 48109-1109; email: FRS@umich.edu

Complex analysis and harmonic analysis to ALEXANDER NAGEL, Department of Mathematics, University of Wisconsin, 480 Lincoln Drive, Madison, WI 53706-1313; email: nagel@math.wisc.edu

Differential geometry and global analysis to LISA C. JEFFREY, Department of Mathematics, University of Toronto, 100 St. George St., Toronto, ON Canada M5S 3G3; email: jeffrey@math.toronto.edu

Dynamical systems and ergodic theory to AMIE WILKINSON, Department of Mathematics, Northwestern University, 2033 Sheridan Road, Evanston, IL 60208-2730; email: transactions@math.northwestern.edu

Functional analysis and operator algebras to DIMITRI SHLYAKHTENKO, Department of Mathematics, University of California, Los Angeles, CA 90095; email: shlyakht@math.ucla.edu

Geometric analysis to WILLIAM P. MINICOZZI II, Department of Mathematics, Johns Hopkins University, 3400 N. Charles St., Baltimore, MD 21218; email: trans@math.jhu.edu

Geometric analysis to MLADEN BESTVINA, Department of Mathematics, University of Utah, 155 South 1400 East, JWB 233, Salt Lake City, Utah 84112-0090; email: bestvina@math.utah.edu

Harmonic analysis, representation theory, and Lie theory to ROBERT J. STANTON, Department of Mathematics, The Ohio State University, 231 West 18th Avenue, Columbus, OH 43210-1174; email: stanton@math.ohio-state.edu

Logic to STEFFEN LEMPP, Department of Mathematics, University of Wisconsin, 480 Lincoln Drive, Madison, Wisconsin 53706-1388; email: lempp@math.wisc.edu

Partial differential equations to GUSTAVO PONCE, Department of Mathematics, South Hall, Room 6607, University of California, Santa Barbara, CA 93106; email: ponce@math.ucsb.edu

Partial differential equations and dynamical systems to PETER POLACIK, School of Mathematics, University of Minnesota, Minneapolis, MN 55455; email: polacik@math.umn.edu

Probability and statistics to KRZYSZTOF BURDZY, Department of Mathematics, University of Washington, Box 354350, Seattle, Washington 98195-4350; email: burdzy@math.washington.edu

Real analysis and partial differential equations to DANIEL TATARU, Department of Mathematics, University of California, Berkeley, Berkeley, CA 94720; email: tataru@math.berkeley.edu

All other communications to the editors should be addressed to the Managing Editor, ROBERT GURALNICK, Department of Mathematics, University of Southern California, Los Angeles, CA 90089-1113; email: guralnic@math.usc.edu.

Titles in This Series

875 **C. Krattenthaler and T. Rivoal,** Hypergéométrie et fonction zêta de Riemann, 2007

874 **Sonia Natale,** Semisolvability of semisimple Hopf algebras of low dimension, 2007

873 **A. J. Duncan,** Exponential genus problems in one-relator products of groups, 2007

872 **Anthony V. Geramita, Tadahito Harima, Juan C. Migliore, and Yong Su Shin,** The Hilbert function of a level algebra, 2007

871 **Pascal Auscher,** On necessary and sufficient conditions for L^p-estimates of Riesz transforms associated to elliptic operators on $\mathbb{R}^n$ and related estimates, 2007

870 **Takuro Mochizuki,** Asymptotic behaviour of tame harmonic bundles and an application to pure twistor D-modules, Part 2, 2007

869 **Takuro Mochizuki,** Asymptotic behaviour of tame harmonic bundles and an application to pure twistor D-modules, Part 1, 2007

868 **Gelu Popescu,** Entropy and multivariable interpolation, 2006

867 **Vilmos Totik,** Metric properties of harmonic measures, 2006

866 **William Craig,** Semigroups underlying first-order logic, 2006

865 **Nathanial P. Brown,** Invariant means and finite representation theory of $C*$-algebras, 2006

864 **John M. Lee,** Fredholm operators and Einstein metrics on conformally compact manifolds, 2006

863 **M. Lübke and A. Teleman,** The Universal Kobayashi-Hitchin correspondence on Hermitian manifolds, 2006

862 **Alberto Canonaco,** The Beilinson complex and canonical rings of irregular surfaces, 2006

861 **Leon A. Takhtajan and Lee-Peng Teo,** Weil-Petersson metric on the universal Teichmüller space, 2006

860 **Thomas M. Fiore,** Pseudo limits, biadjoints and pseudo algebras: Categorical foundations of conformal field theory, 2006

859 **N. Arcozzi, R. Rochberg, and E. Sawyer,** Carleson measures and interpolating sequences for Besov spaces on complex balls, 2006

858 **Enrico Valdinoci, Berardino Sciunzi, and Vasile Ovidiu Savin,** Flat level set regularity of p-Laplace phase transitions, 2006

857 **Donatella Danielli, Nocola Garofalo, and Duy-Minh Nhieu,** Non-doubling Ahlfors measures, perimeter measures, and the characterization of the trace spaces of Sobolev functions in Carnot-Carathéodory spaces, 2006

856 **Vladimir Bolotnikov and Harry Dym,** On boundary interpolation for matrix valued Schur functions, 2006

855 **Yevgenia Kashina, Yorck Sommerhäuser, and Yongchang Zhu,** On higher Frobenius-Schur indicators, 2006

854 **Noam Greenberg,** The role of true finiteness in the admissible recursively enumerable degrees, 2006

853 **Joachim Krieger,** Stability of spherically symmetric wave maps, 2006

852 **Viorel Barbu, Irena Lasiecka, and Roberto Triggiani,** Tangential boundary stabilization of Navier-Stokes equations, 2006

851 **Jie Wu,** On maps from loop suspensions to loop spaces and the shuffle relations on the Cohen groups, 2006

850 **Siegfried Echterhoff, S. Kaliszewski, John Quigg, and Iain Raeburn,** A categorical approach to imprimitivity theorems for C^*-dynamical systems, 2006

849 **Katsuhiko Kuribayashi, Mamoru Mimura, and Tetsu Nishimoto,** Twisted tensor products related to the cohomology of the classifying spaces of loop groups, 2006

848 **Bob Oliver,** Equivalences of classifying spaces completed at the prime two, 2006

TITLES IN THIS SERIES

847 **Eric T. Sawyer and Richard L. Wheeden,** Hölder continuity of weak solutions to subelliptic equations with rough coefficients, 2006

846 **Victor Beresnevich, Detta Dickinson, and Sanju Velani,** Measure theoretic laws for lim-sup sets, 2006

845 **Ehud Friedgut, Vojtech Rödl, Andrzej Ruciński, and Prasad V. Tetali,** A Sharp threshold for random graphs with a monochromatic triangle in every edge coloring, 2006

844 **Amadeu Delshams, Rafael de la Llave, and Tere M. Seara,** A geometric mechanism for diffusion in Hamiltonian systems overcoming the large gap problem: Heuristics and rigorous verification on a model, 2006

843 **Denis V. Osin,** Relatively hyperbolic groups: Intrinsic geometry, algebraic properties, and algorithmic problems, 2006

842 **David P. Blecher and Vrej Zarikian,** The calculus of one-sided M-ideals and multipliers in operator spaces, 2006

841 **Enrique Artal Bartolo, Pierrette Cassou-Noguès, Ignacio Luengo, and Alejandro Melle Hernández,** Quasi-ordinary power series and their zeta functions, 2005

840 **Sławomir Kołodziej,** The complex Monge-Ampère equation and pluripotential theory, 2005

839 **Mihai Ciucu,** A random tiling model for two dimensional electrostatics, 2005

838 **V. Jurdjevic,** Integrable Hamiltonian systems on complex Lie groups, 2005

837 **Joseph A. Ball and Victor Vinnikov,** Lax-Phillips scattering and conservative linear systems: A Cuntz-algebra multidimensional setting, 2005

836 **H. G. Dales and A. T.-M. Lau,** The second duals of Beurling algebras, 2005

835 **Kiyoshi Igusa,** Higher complex torsion and the framing principle, 2005

834 **Kenîchi Ohshika,** Kleinian groups which are limits of geometrically finite groups, 2005

833 **Greg Hjorth and Alexander S. Kechris,** Rigidity theorems for actions of product groups and countable Borel equivalence relations, 2005

832 **Lee Klingler and Lawrence S. Levy,** Representation type of commutative Noetherian rings III: Global wildness and tameness, 2005

831 **K. R. Goodearl and F. Wehrung,** The complete dimension theory of partially ordered systems with equivalence and orthogonality, 2005

830 **Jason Fulman, Peter M. Neumann, and Cheryl E. Praeger,** A generating function approach to the enumeration of matrices in classical groups over finite fields, 2005

829 **S. G. Bobkov and B. Zegarlinski,** Entropy bounds and isoperimetry, 2005

828 **Joel Berman and Paweł M. Idziak,** Generative complexity in algebra, 2005

827 **Trevor A. Welsh,** Fermionic expressions for minimal model Virasoro characters, 2005

826 **Guy Métivier and Kevin Zumbrun,** Large viscous boundary layers for noncharacteristic nonlinear hyperbolic problems, 2005

825 **Yaozhong Hu,** Integral transformations and anticipative calculus for fractional Brownian motions, 2005

824 **Luen-Chau Li and Serge Parmentier,** On dynamical Poisson groupoids I, 2005

823 **Claus Mokler,** An analogue of a reductive algebraic monoid whose unit group is a Kac-Moody group, 2005

822 **Stefano Pigola, Marco Rigoli, and Alberto G. Setti,** Maximum principles on Riemannian manifolds and applications, 2005

For a complete list of titles in this series, visit the
AMS Bookstore at **www.ams.org/bookstore/**.